Worms
Eat My
Garbage

**HOW TO SET UP AND MAINTAIN
A WORM COMPOSTING SYSTEM**

By Mary Appelhof and Joanne Olszewski

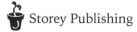
Storey Publishing

The mission of Storey Publishing is to serve our customers by publishing practical information that encourages personal independence in harmony with the environment.

Edited by Corey Cusson and Carleen Madigan
Art direction and book design by Jeff Stiefel
Text production by Jennifer Jepson Smith
Indexed by Nancy D. Wood
Cover and interior illustrations by © Phil Hackett, except for page 114,
    adapted from Dindal 1971, *Ecology of Compost*
Plan drawings pages 21, 24, 25, and 27 by Ilona Sherratt

© 2017 by Joanne Olszewski
Previous editions of this book were published under the same title by
    Flower Press.

Storey Publishing
210 MASS MoCA Way
North Adams, MA 01247
storey.com

Printed in the United States by Versa Press.
10 9 8 7 6 5 4

LIBRARY OF CONGRESS CATALOGING-IN-PUBLICATION DATA

Names: Appelhof, Mary, author. | Olszewski, Joanne, author.
Title: Worms eat my garbage / by Mary Appelhof and Joanne Olszewski.
Description: 35th anniversary edition. | North Adams, MA : Storey Publishing,
    2017. | Includes bibliographical references and index.
Identifiers: LCCN 2017034444 (print) | LCCN 2017047000 (ebook)
    | ISBN 9781612129488 (ebook) | ISBN 9781612129471 (pbk. : alk. paper)
Subjects: LCSH: Earthworm culture. | Earthworms. | Vermicomposting.
    | Refuse and refuse disposal.
Classification: LCC SF597.E3 (ebook) | LCC SF597.E3 A67 2017 (print)
    | DDC 639/.75—dc23
LC record available at https://lccn.loc.gov/2017034444

Contents

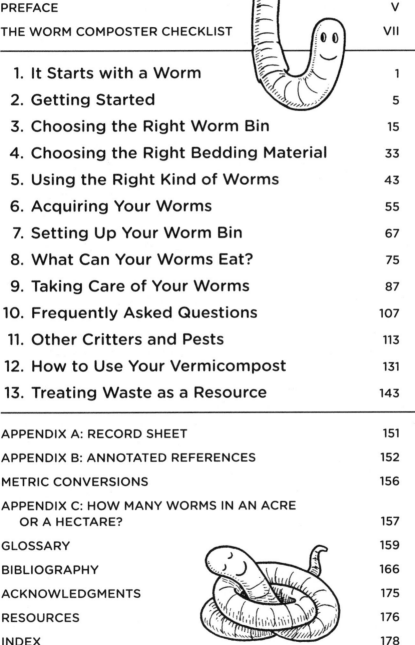

Foreword

In 2001, I called Mary Appelhof and told her that I wanted to write a natural history of earthworms. "At first I wasn't sure anyone would want to read a book about worms," I said, "but I'm finding out that everybody has an earthworm story."

"Oh, they do!" she said, and proceeded to tell me hers. She built her first worm bin in 1972, and published a brochure about worm composting a year later. "I sold it to anyone who would send a quarter and a self-addressed, stamped envelope," she said.

"Did it sell?" I asked.

"It was a hit!" she said. "I did a survey back then on people's attitudes towards using worms for home garbage disposal. Only 25% of the people who responded to the survey said they couldn't stand the idea. I figured that meant 75% were willing to consider it."

That was Mary: the eternal optimist, the tireless evangelist. Anytime she saw an opportunity to promote earthworms, she took it. She taught classes at any school or nursery that would have her, she spoke at conferences, produced educational videos, and began answering to the name "Worm Woman."

She published *Worms Eat My Garbage* in 1982, when self-publishing a book meant stacking thousands of copies in the garage and peddling them out of the trunk of the car. It was the only book about worm composting on the market at the time, and remains the best: it's thorough, well-researched, and entertaining. To everyone's astonishment but Mary's, the book sold 100,000 copies — but not overnight.

"It only took me twenty years!" she said. "When I started, I envisioned huge piles of garbage and huge quantities of worms. I didn't have the wherewithal to make that happen, but I did know how to get worm composting going one household at a time. So that's what I did."

The book you hold in your hands is nothing less than Mary Appelhof's prescription for saving the world — in your own backyard. It's now been twenty years since I started my first worm bin. In that time they've proven to be surprisingly good pets: clean, industrious, self-sufficient, and always up for the job of devouring compost and enriching the soil.

If you're embarking on your first adventure with earthworms, I congratulate you on your decision and promise that before long, you're going to have your own earthworm story to tell.

AMY STEWART, author of *The Earth Moved: On the Remarkable Achievements of Earthworms*

Preface

I met Mary Appelhof in 1982. Her first gift to me was a 1-2-3 Worm Box, complete with worms. At work, I used the paper shredder to shred newspaper. Adding the paper and vegetable scraps to the bin, I then placed it under a bench in my greenhouse and totally forgot about it. About a month later, Mary came to visit and checked up on the worms. Imagine my surprise when she pulled back the bedding to find one of the most prolific masses of worms she had ever seen. I have been worm composting since that time, and I occasionally forget about these amazing creatures, yet they continue to perform. Today, I have two bins. The large outdoor one I made from cedar pickets, and the second, small indoor bin is Mary's Worm-A-Way that I use for demonstrations.

Mary passed away in 2005 and the first two editions of *Worms Eat My Garbage* were written in her own words. Many times she used the first person "I" to explain her ideas and reasoning. I have chosen to leave her words intact. In the new sections of the book, I also used the first person to express myself. Although it could be confusing to have two authors each use "I" in one book, rest assured that I agreed with Mary on her ideas and I believe she would agree with mine.

This third edition continues to answer the most common vermicomposting questions while updating and expanding on the scientific data. Mary had a gift for taking difficult information and presenting it in a simple fashion. As a former middle school science teacher, I have attempted to continue that practice. This update also answers questions Mary asked in her previous editions. Much has happened in composting, vermicomposting, and recycling in the last two decades. It was Mary in 1993 who coined the term "worm workers." This revision brings us up-to-date on what worm workers are doing and hopefully inspires you to become a worm worker as well. After all, if worms eat my garbage, they will eat yours, too.

— Joanne Olszewski, Worm Woman 2.0
2017

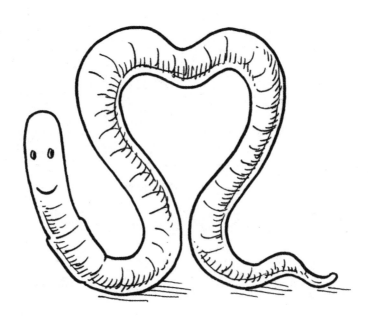

The Worm Composter Checklist

1   Read *Worms Eat My Garbage*.

2   Weigh organic kitchen waste for two to three weeks to get the average amount produced in your household so that you can determine the right size bin for the amount of waste you produce.

3   Determine the quantity of worms you need and order worms.

4   Purchase a bin or select the size of container required and assemble the materials.

5   Determine what types of bedding are available, and either order materials or scrounge.

6   Build or assemble your bin.

7   Prepare the bedding. If using manure, prepare it at least two days before the arrival of your worms.

8   Add worms to the bedding.

9   Bury your food waste in the bedding.

10   Check moisture levels periodically; look for cocoons and young worms.

11   Harvest worms and prepare new bedding.

12   Use the vermicompost or worm castings on houseplants or in your garden.

Earthworms in nature play an important role in recycling organic nutrients from dead tissues back into living organisms. They do this without fanfare; rarely does anyone see them perform their tasks.

If you decide to use composting worms to process your own organic kitchen waste, you will see them at work. You will see mounds of disagreeable material converted noiselessly, with almost no odor, to material you can use directly on your house-plants and in your garden. You will enjoy healthier-looking plants, better-tasting vegetables, and money in the bank. You will spend less on fertilizers and trash disposal. Some of you will be serving fish for dinner — caught using "your" worms as bait. Hopefully, you'll also gain a better appreciation of the intricate balance and interdependencies in nature. You will be treading more gently upon the Earth.

As your gardens are enriched, so is your life and mine. You will have joined the worm-working adventurers who say, "Worms eat my garbage." Isn't that a grand beginning to a task that needs to start somewhere? You, personally, can make it happen.

WHAT TO CALL YOUR SETUP

Some people use the term "home vermicomposting system" because it sounds more sophisticated than "worm bin." They are right on both counts; it is sophisticated, and it is a system. The system consists of five interdependent parts:

Physical structure: A box or container

Biological organisms: The worms and their associates

Controlled environment: Temperature, moisture, acidity, ventilation

Maintenance procedures: Preparing bedding, burying kitchen and food waste, separating worms from their castings

Production procedures: Making use of the castings (worm manure)

I hesitate to use "home vermicomposting system" exclusively because the term itself might frighten away those who feel more comfortable with "worm bin." It sounds less intimidating to suggest taking a plastic container or wooden box and putting holes in it to provide a source of air. Then add damp bedding and worms, bury food waste, harvest worms, and set up fresh bedding as necessary. If calling this system a worm bin encourages you to try the technique, no other term is better.

On the other hand, the system truly is complex. Much more can be learned about it: Just what is going on in that bin? Are the worms really eating the food waste, or are they eating the microbes, bacteria, protozoa, molds, and fungi that are breaking down the food waste? Can conditions become too acidic? How can you tell? What kinds of food might cause overacidity? When is the best time to harvest if you want the greatest number of worms for the least amount of time and effort? When do you harvest to get the best castings? What is the best size of container to bury a given quantity of food waste?

If you like to compare notes, for example, about ideal temperature for worm cocoon production or about acceptable levels of anaerobiosis (absence of oxygen), you might want to say you have a home vermicomposting system. I have such a system. I know a great deal about vermicomposting, but I have a lot more to learn. For myself, I just say, "Worms eat my garbage. Wanna see my worm bin?"

When deciding what kind of worm bin to get and where to put it, consider both the worms' needs and your own.

That may sound elementary, but I've learned from experience that there are a few basics you should think about in advance. Adjusting your thinking early will help determine later how successful and enjoyable a worm bin is for you.

To meet your needs, a home vermicomposting system will have to measure up to your expectations, provide a convenient method for converting organic waste to a usable end product, and satisfy your idea of suitable aesthetics. Potential end products are a supply of worms for fishing, worm castings for plants, or vermicompost for use in your garden. **If you do use worms for fishing, it is important that you are aware of their proper disposal.** (See chapter 5 for more information on worms as an invasive species.)

WHAT TO EXPECT

The effectiveness of your vermicomposting system will depend partly upon your expectations and partly upon your behavior. You can reasonably expect to bury a large portion of your bio-degradable kitchen or food waste in a properly prepared worm bin, check it occasionally, make judgments about what must be done, then harvest worms and vermicompost or worm castings after a period of several months. You cannot expect to merely

Compost or Castings?

Vermicompost is a more general term than worm castings. A casting is manure, the material deposited from the anus after it has moved through the digestive tract of a worm. Vermicompost contains worm castings but also consists of partially decomposed bedding and organic waste with recognizable fragments of plants, food, and bedding. Worms of all ages, cocoons, and associated organisms may be found in vermicompost. If a worm bin is left untended for six months or so, worms will eat all of the bedding and organic waste, depositing castings as they do so. In time, they will have reingested the materials a number of times. Then the entire contents will have been fully converted to vermicast, which is completely worm-worked and reworked material with a fine, smooth texture. Vermicast is considered over-worked and has probably lost nutrients. At that stage, since no food remains for the worms, most worms will die and be decomposed by the other organisms in the worm bin. The few worms that live will be small, inactive, and undernourished.

dump all the trash from your kitchen into a worm bin, add some worms, and come back in only two weeks to collect quantities of fine, dark worm castings to sprinkle on your houseplants. Either revise this unreasonable expectation to something more realistic, perhaps using the ideas below, or don't even set up a system.

Your Goals

The difference between reasonable and unreasonable expectations has to do with the kind of food waste you bury, the environment you provide for the worms, the length of time you are willing to wait to observe changes, and the character of the end products. Success isn't that difficult when you know what you want. Guidelines to help you make reasonable judgments about maintaining your worm bin effectively depend on your goals.

GOAL: EXTRA WORMS FOR FISHING

MAINTENANCE LEVEL: HIGH

Some of you will want to produce more worms than you started with so that you have a ready supply for fishing. Expect to harvest your bin every two to three months, transfer worms to fresh bedding, and accept vermicompost that is less finished than if you were to leave the worms in their original bedding longer.

Producing more worms for fishing bait means having less finished vermicompost.

GOAL: FINISHED WORM CASTINGS FOR PLANTS

MAINTENANCE LEVEL: LOW

Those who prefer to obtain castings in the most finished form have the advantage of extremely low maintenance. You will bury food waste in your worm bin over a four-month period, then leave it alone. You won't have to feed or water the worms for the next few months, letting the entire culture proceed at its own pace. The worms will produce castings continuously as they eat the bedding and food waste. The disadvantage of this program is that as the proportion of castings increases, wastes that are toxic to the worms accumulate, and the environment for worms becomes less healthy. They get smaller and stop reproducing, and many die. As you wait for the worms to convert all the bedding and food waste to castings, you will have to deposit fresh batches of food waste elsewhere — perhaps in a compost pile or in a second worm bin.

In time, your worm bin will provide a quantity of fine castings to give you a homogeneous, nutrient-rich potting soil. If you choose this goal, you can alternate from one bin to another. Or you may have to purchase worms every fall when you set up your worm bin. This low-maintenance program enables you to vermicompost inside during the cold winter months, compost outside in the traditional manner when the weather warms, and have

Worm castings enrich the soil for your garden or ordinary houseplants.

finished worm castings from your indoor worm bin sometime during summer.

A nickname for this maintenance approach might be "the lazy person's technique." However, if you use two bins, it's actually a pretty efficient system; "the smart person's technique" fits the situation better.

GOAL: CONTINUOUS WORM SUPPLY PLUS VERMICOMPOST
MAINTENANCE LEVEL: MEDIUM

Middle-of-the-roaders can opt for a program that requires just enough maintenance to keep the worms healthy. You will harvest fewer worm castings, but you should still have ample quantities of vermicompost to use on your houseplants and garden and enough worms to set up your bin again. About every four months, you will prepare fresh bedding and select one of several techniques for separating worms from vermicompost. These techniques are described in chapter 9.

It should be apparent that the effectiveness of your home vermicomposting system depends as much upon you as it depends upon the worms.

Location, Location, Location

The convenience of your home vermicomposting system is directly related to its location. There are various possibilities. For some of you, locating your worm bin will depend considerably on what it looks like. A homemade bin may be practical but not necessarily beautiful. If it is a custom-made unit of laminated maple with sturdy legs on ball rollers and looks like a piece out of a Nieman Marcus catalog, you will want it where you can show it off most readily. Constructing your own worm bin is, of course, always an option; DIY instructions abound online, and classes may be offered at some organizations (see Resources, page 176, for more details).

More realistically, before you decide where to locate your worm bin, (1) determine how large your unit must be to process your kitchen waste (see page 17 for more on this), (2) assess the space you have available, and (3) determine whether you want it to be merely functional and out of the way or the center of attention. To state this another way: How many guests do you want tramping through your basement? Or how many guests can deal with sitting on a window-seat worm bin in the dining room? From my experiences, I can guarantee you, until worm bins are common, almost everyone who visits is going to want to see yours!

THE KITCHEN

Since food preparation is done in the kitchen, the most convenient location for a worm bin might be there, too. One of my friends has a worm bin on top of his dishwasher with a cutting board serving as a lid. When he is through chopping cabbage, celery, or whatever, he just slides the top back and scrapes the waste into his worm bin. You can't beat that for convenience!

THE PATIO

An outdoor patio off the kitchen is an excellent location for a home vermicomposting unit if it will be out of direct sun during the summer months. It is close to the origin of food waste, is near a water supply for maintaining proper moisture, and has plenty of ventilation. Just as you can expect to get dirt on the floor when you mix potting soils to repot plants, know that the periodic maintenance of separating worms from vermicompost can get messy. Doing it on the patio will keep the dirt outside. In climates where freezing temperatures are a problem, insulation and supplemental heat can keep the worms going. I describe some of these adaptations in chapter 9.

THE BALCONY

Apartment dwellers often have limited space. In warmer climates, many people living in apartments find that their balcony accommodates a worm bin and a few container plants. They like the appearance of the plants and feel good about doing something useful with their food waste. The plants give them a place to put the vermicompost produced in the worm bin.

THE GARAGE

A well-ventilated garage would be a satisfactory location for a worm bin if it blocks freezing temperatures in your production area. It also will provide shade during hot weather.

BASEMENT

Locating a worm bin in a basement, if you have one, has the advantage of keeping it out of the way. If problems develop — there might be short-term odors or occasional fruit flies — the worm bin is not in the immediate living quarters. You might find it inconvenient to always go downstairs whenever you want to bury food waste, however. Basements do meet the worms' temperature needs, being cooler in summer and rarely freezing in winter. Since they are out of sight and not in the way for most people, many worm bins, including mine, are located in a basement.

YOUR WORMS' NEEDS

To make the worms happy, you'll need to think about temperature, moisture, acidity, and ventilation. Equally as important, to make yourself happy, you'll want to consider your expectations, sense of convenience, and aesthetic preferences.

Keeping Them Warm

You should be using the redworm, *Eisenia fetida*; the red tiger, *E. andrei*; or a combination of both. These are best for home composting, for reasons that I will discuss later (see page 46). They feed most rapidly and convert waste best at temperatures between 59 and 77°F (15 and 25°C), with an optimal temperature of 77°F. They can work their way through food waste in a basement bin with temperatures as low as 50°F (10°C), but below this they reduce their feeding. Ultimately, these worms can survive in temperatures between 32 and 95°F (0 and 35°C). Below 39°F (3.9°C), cocoon production and development of young ceases.

Redworms have successfully weathered cold northern winters in pits dug into the ground that were covered with manure, straw, and leaves to provide heat, food, and insulation. The problem with an outdoor pit for winter food waste disposal is disturbing the protective snow covering to bury food waste. When the temperature drops to zero, such protection saves the worms, but your food waste also piles up!

Temperatures over 86°F (30°C) promote microbial activities that can consume oxygen that the worms need. The temperature in moist bedding is generally lower than the surrounding air because evaporation of moisture from the bedding in a well-ventilated place has a cooling effect. Locations that could get too hot include a poorly ventilated, overheated attic; outside under a hot sun; and in a greenhouse in higher elevations.

Keeping Them Happy

MAINTAINING MOISTURE. All worms need moisture. They "breathe" through their skin, which must be moist for the exchange of air and the excretion of waste to take place. You can add water to dry bedding when necessary. Too much moisture, present as water standing in the bin, can reduce available oxygen and cause worms to "drown." This can be a problem in plastic bins, a point I will discuss in more detail in chapter 9. Location is another factor with excess moisture. Place your worm bin where there is no danger of natural flooding, which could also drown the worms.

MAINTAINING ACIDITY. Redworms can tolerate a fairly wide range of acidity in their environment, but slightly acid conditions are best. The 14-point scale for determining degree of acidity is called pH. The most acid reading would be pH 1; the most alkaline reading would be pH 14. Neutral is pH 7, meaning that the medium is neither acid nor alkaline. A wide range from pH 5 to pH 9 is suitable for redworms. In a worm bin with a pH of 4, you may find worms dying or trying to escape from the excessive acidity in their environment. Providing too much acidic food would be rather like pouring a bottle of vinegar into a worm bin — not a good idea! The most accurate method for testing the pH of your worm bin is to use a pH meter. Simply place the meter in at least three different places in the bin to check the pH level. If you don't have a meter, then use the smell test. If it smells bad, it might be a pH problem. I personally like using the meter because it is a simple, accurate, economical way to determine the pH.

MAINTAINING VENTILATION. Worms use oxygen in their bodily processes, producing carbon dioxide, just as we do. It is important that you allow air to circulate around your container as a structural unit. Wrapping it in a plastic bag, for example, might be tidy, but the worms would quickly smother. For more on ventilation, see chapter 3, page 15.

A variety of containers make satisfactory worm bins. These range from commercially available vermicomposting units to containers you adapt or bins you build yourself.

Regardless of your choice, aeration is an essential function of the container's controlled environment. The greatest concern people express when they hear about placing kitchen waste in a container to be kept inside the home is "But won't it smell?" The answer: "Not too badly if it is properly set up and maintained." That is, when your bin is set up so that it has good airflow.

ENSURING GOOD AIRFLOW

We are trying to create an *aerobic* environment, one in which oxygen is present throughout the bedding. Oxygen is necessary not only for the worms but for the millions of aerobic microorganisms that also break down the food waste. As the worms and microorganisms consume the organic material, oxygen from the air combines with carbon in the food waste in a process that releases energy. They give off carbon

The Secret to an Odor-Free Worm Bin

In the absence of oxygen — a state known as anaero-biosis — worms become unhealthy and may die as anaerobic microor-ganisms break down the food waste. These microorganisms live and reproduce only when no oxygen is present. As they break down waste, they produce gases that have foul-smelling, disagreeable odors. You can tell that anaerobes are working when you remove the lid from a smelly garbage can. The secret to having an odor-free worm bin is to have oxygen available throughout the bedding so that both the worms and different kinds of microor-ganisms can break down the waste aerobically.

dioxide, water, and other plant nutrients as waste products. The advantage to our senses is that neither carbon dioxide nor water smells.

LOTS OF SURFACE AREA. Given bins of different shapes but equal volume, the one with greater surface area allows more air to contact the bedding and provides more surface on which to place waste. Some commercial units use a system with several layers to support worms and bed-ding. The worms work from the bottom up through a mesh screen into the layer above. Such stack-ing units allow the unit to take up less floor space but still provide much surface area for aeration.

If your worm bin is a single layer, it should be shallow — usually no deeper than 12" to 18". The reason for this shallowness is that redworms tend to feed upward, nibbling from beneath the material on the surface. More surface, more nibbling. Bedding can pack down in a deep con-tainer. Such compression pushes the air out of the bottom layers, and the consequent development of foul-smelling *anaerobic* condi-tions (meaning without available oxygen) is more likely (see The Secret to an Odor-Free Worm Bin, left).

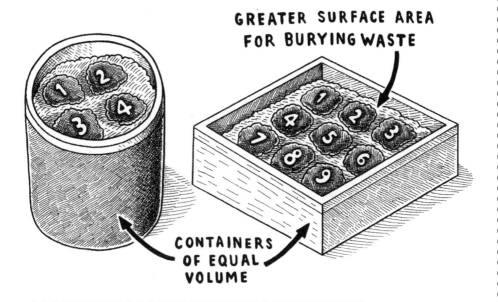

GREATER SURFACE AREA FOR BURYING WASTE

CONTAINERS OF EQUAL VOLUME

More surface area means you can bury waste more frequently.

VENTILATION. Because redworms need lots of oxygen, whatever the shape of their bin, their container must have holes in the sides, top, or bottom to let air in but keep flies out. Sometimes a mesh screen is used to cover the holes. Or a rectangle is cut out of the bottom, side, or lid and covered with screen to provide more opportunity for air to get to the bedding. Various sizes and styles of louvered vents also permit air exchange.

Choosing the Right Size Bin

How do you decide how big your home vermicomposting unit should be? First, you need some idea of how much organic kitchen waste you produce. I recommend that you keep track of how many pounds you throw away each week. Is it 5 pounds (2.3 kg)? Or 10 pounds (4.5 kg)? Many factors affect this. How many individuals live in your household? Are you vegetarians who generate more worm food than meat eaters? How often do you eat out? Do you use prepared mixes or start meals from scratch? How often do you throw away leftovers and spoiled food? The amount of food for worms also varies depending on the menu, visitors, and vegetables in season.

EXAMPLE 1: A COUPLE OF VEGGIE-LOVERS

My household of two adults produces about 4 pounds (1.8 kg) of worm food per week. It consists of such waste as potato peels, citrus rinds, outer leaves of lettuce and cabbage, tea bags or herb tea leaves, moldy leftovers, plate scrapings, cucumber peelings, pulverized eggshells, and onion skins. We often eat lunch at home and may eat dinner out two or three times a week. Based upon 1 square foot of surface for each pound of food (0.5 kilogram) waste per week, we need a bin with 4 square feet of surface area. A bin measuring 12" × 24" × 24" would give us the 4 square feet needed for our 4 pounds (1.8 kg) of worm food.

EXAMPLE 2: A FAMILY OF PICKY EATERS

A single-parent household with two children, where the parent loves to cook, has anywhere from 1.75 to 12 pounds (0.8 kg to 5.4 kg) of waste to feed worms per week. Their average over a 14-week period is 5.2 pounds (2.4 kg) per week. They use a bin that is 12" × 24" × 36". This size still approximates 1 square foot for each pound of food waste per week (0.5 kilogram).

Selecting the Structural Material

Common materials for worm bins are wood and plastic. Wooden bins "breathe" more than plastic, but they deteriorate more quickly because the wood is damp all the time. In dry conditions, the bedding in wooden bins can dry out, so water must be added when needed. Just the wood in a wooden bin is heavy; by the time you add bedding, water, worms, and food waste, it gets very heavy. Moving it is out of the question.

Plastic bins require more holes for aeration than wooden bins and tend to accumulate excess moisture. Plastic containers initially used for other purposes are often readily available, however, and are easy to adapt by drilling holes in them. Just

CAUTION: If you want to improvise with containers on hand, be sure that the one you select has not been used to store chemicals, such as pesticides, which may kill the worms. Some worm growers suggest that you scrub new plastic containers with a strong detergent, then rinse them carefully before you grow worms in them.

remember, whether you purchase a commercial unit or build your own, a closed bin with no provision for aeration will produce as many odors as a closed garbage can, and you'll kill your worms before they even get a chance to eat your waste.

WOODEN WORM BINS

How long will a wooden box last? Used continuously, without ever letting the box dry out, unfinished wooden boxes might last two to three years. Longevity can be increased by letting the box dry out for several days between setups. Some people alternate between two boxes for more convenient maintenance. A good finish, such as polyurethane varnish, epoxy, or other waterproofing material, that seals all edges should extend box life considerably.

CAUTION: Pressure-treated lumber is not recommended for worm bin interiors. Up until 2003, treated wood (called CCA wood) contained copper, chromium, and arsenic. Manufacturers voluntarily discontinued using CCA wood for use by residential homeowners because of potential health risks. Two common wood preservatives used today are ACQ (alkaline copper quaternary) and CA (copper azole). Another treatment contains micronized copper preservatives. Although newer products may be less toxic than older CCA wood, I do not recommend using any pressure-treated wood in the garden or in worm bins.

A simple wooden box made of boards, plywood, recycled wood, or an old drawer works as a worm bin if you drill several 1" holes in the sides about 2" to 3" from the bottom and then another row along the top of the other side. Two very common sizes for wooden worm bins are 12" × 24" × 36" and 24" × 24" × 8". The first of these foot-measured lengths is sometimes called a 1-2-3 Worm Box. It is large enough to handle an average of 6 to 7 pounds (3 kg) of waste per week, or the food waste expected from a family of four to six people. These units require only a hammer and a drill with a 1" bit; constructing larger bins for multiple purposes involves a longer list of appropriate tools.

Note that the CDX plywood, mentioned below, is exterior grade (the C and D denote how sanded the plywood is, and the X means exterior). Scrap lumber or #2 pine boards can be substituted. Ardox nails have a spiral shank that increases their holding power, particularly important for wood that is alternately wet and dry. Deck screws are easier to use than nails, hold together better than nails, and allow for easier changes. Also, please note that a dimensional 2×4 board will actually measure $1\frac{1}{2}$" × $3\frac{1}{2}$".

The chart below shows how to interlock the corners for greater strength. Once the sides are screwed together with about four screws per side, secure the sides to the bottom plywood sheet using five to seven screws per side. Drill six 1" holes in each side: one row along the bottom as shown, and another row along the top of the other side. These allow aeration. Having a row along the top of one side permits air to escape if the temperature gets too high in the bin. Interestingly, the worms don't usually crawl out of the holes. Stapling screen over the holes on the inside of the bin will prevent material from falling out and flies from getting in, although this step is not critical in most cases.

1-2-3 Worm Box Materials

Quantity	Item
2	$\frac{5}{8}$" × $35\frac{3}{8}$" × 1' pieces CDX plywood
2	$\frac{5}{8}$" × $23\frac{3}{8}$" × 1' pieces CDX plywood
1	$\frac{5}{8}$" × 2' × 3' piece CDX plywood
38	2" or $1\frac{1}{2}$" Ardox nails or screws

1-2-3 WORM BOX

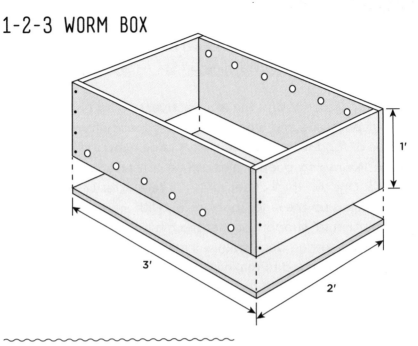

The construction diagram for the 1-2-3 Worm Box includes a side overlap on the corners for strength and holes in the sides for aeration.

TWO-PERSON BIN

Substitute woods may be used for this bin, too. Screw the sides together, overlapping the corners as shown in the illustration above. Secure the bottom to the sides using about five screws per side. Drill three 1" holes about 2" from the bottom of one side and the top of the opposite side for aeration.

Two-Person Bin Materials

Quantity	Item
4	$^5/_8$" × 23$^3/_8$" × 8" pieces CDX plywood
1	$^5/_8$" × 2' × 2' piece CDX plywood
36	2" or 1$^1/_2$" Ardox nails or deck screws

PATIO BENCH WORM BIN

A more elaborate but effective design is the Patio Bench Worm Bin, adapted from a design developed by Seattle Tilth, an urban gardening organization in Washington State (see Resources, page 176).

The Patio Bench Worm Bin is attractive, serving both as a waste disposal site and a bench. With a capacity from 8 to 16 pounds of food waste per week, it is large enough for most families of four to six people, and can be constructed in nine easy steps. One ½"-thick sheet of 4- × 8-feet exterior-grade plywood makes up the sides, bottom, and lid. Framing lumber provides structural support so that the bin is strong enough to sit upon. Water-based deck stain will make it more attractive and help seal the wood to make it last longer. Another option for preserving the wood is to coat the wood inside and out with vegetable oil.

TOOLS NEEDED

- Portable electric saw or hand crosscut saw; electric drill with variable speed or an impact driver; driver bits to match screw heads; ³/₈" drill bit (I highly recommend a longer driver bit to make the work easier.)

- Tape measure; hammer; sawhorses; straightedge; wood chisel; wood glue

- Paintbrush for application of vegetable oil or stain (You can also apply vegetable oil with a cloth.)

- Ear and eye protection

- Respirator or face mask

Plywood Patio Bench Worm Bin Materials

Quantity	Item
3	8' 2×4
1	6' 2×4
1	$1/2$" × 4' × 8' piece of CDX plywood
12	$3^1/2$" star deck screws (to fasten plywood into 2×4 stock)
80	$1^1/4$" star deck screws
2	3" T hinges
1	4' length of $3/8$" polypropylene rope
1	quart water-based deck stain
4	$1/2$"-wide screw eyes

Cedar Patio Bench Worm Bin Materials

Quantity	Item
20	6' lengths of 1"×6" dog-ear cedar pickets

Materials are the same for the Cedar Patio Bench Worm Bin as the Plywood Patio Bench Worm Bin with one exception: use dog-ear cedar pickets in place of plywood.

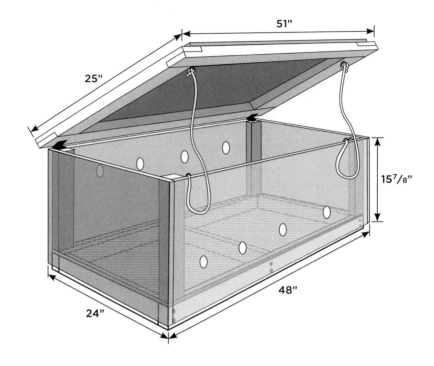

PLYWOOD CUTTING DIAGRAM

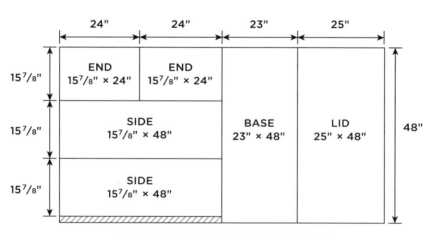

4' × 8' SHEET OF HALF-INCH PLYWOOD

FRAMING LUMBER CUTTING DIAGRAM

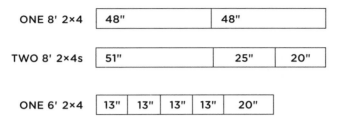

ONE 8' 2×4	48"	48"

TWO 8' 2×4s	51"	25"	20"

ONE 6' 2×4	13"	13"	13"	13"	20"

BASE FRAME DIAGRAM

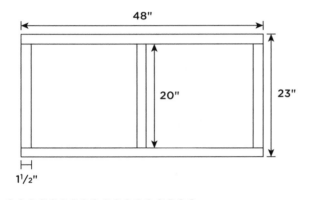

48"

20"

23"

1¹/₂"

The base frame has each cut piece of 2×4 on edge.

STEP 1. Cut the plywood as shown in the Plywood Cutting Diagram opposite. Measure a full 48" from the left for the first cut, and make the base a full 23" × 48". The width of the saw cuts will make the final measurement less than a full 25" for the lid, but this won't matter.

STEP 2. Cut the framing lumber as shown in the Framing Lumber Cutting Diagram above.

STEP 3. Assemble the 2×4 base. Place the three 20" and two 48" base sides on edge. Predrill and use two 3¹/₂" screws to secure the base sides to the ends and center spacers as in the Base Frame Diagram above. Then fasten the 23" × 48" plywood to the base. This sheet will be the floor of your worm bin.

STEP 4. Secure the 2×4 corner supports to the plywood sides. Place two corner supports flat side down on a large, flat surface. Lay a 15^7/$_8$" × 48" side piece on the corner supports, line up the top and side edges, and fasten the plywood to each corner support with three or four 1^1/$_2$" screws. Repeat for the second side.

STEP 5. Fasten the ends to the sides to make the box. With a helper, place the sides against the base and secure the plywood ends to the side panels, screwing into the corner supports. These supports provide a solid surface for fastening and strengthening the box. Note that the side panels do not extend all the way to the ground. This helps to prevent moisture from seeping up into the edges of the plywood from the ground. Screw the box around the base, using a fastener every 3" to 4".

STEP 6. Drill ventilation holes in the side panels. Drill four to eight 1" holes about 3" from the floor on one side of the bin. Drill a similar number of holes about 3" from the top of the other side of the bin to help with ventilation. Warm air can exit from the top, drawing cooler air in from the bottom. Installation of ventilation louvers (soffit vents) in these holes after staining will dress up the appearance of the bin and help keep insects or rodents from entering.

STEP 7. Assemble the lid. The plywood lid is supported by a 2×4 frame with lap joints in the corners. To make the lap joints, set the saw to cut ^{3}/$_4$" deep, which is half the thickness of a 2×4. Taking the 51"-long and 25"-long boards, measure 3^1/$_2$" from each end. Make shallow cuts every ^{1}/$_2$" up to the 3^1/$_2$" mark. Then use a sharp wood chisel to knock out the extra wood. Be consistent with which half of each end is cut out — for example, you may cut the top half out of the ends, the bottom half out of the sides. Repeat this eight times, once for each end of the four boards. Glue and secure the lap joints together with 1^1/$_2$" screws. Once the glue has set, center the 25" × 48" plywood lid on the frame and fasten it with 1^1/$_2$" fasteners every 6".

STEP 8. Apply the hinges. Attach the hinges to the back edge of the bin centered on the 2×4 corner blocks. Attach the hinges with screws after you have drilled pilot holes.

STEP 9. Provide lid support. Use polypropylene rope to keep the lid from falling backward and to reduce strain on the hinges. Polypropylene will last longer without deterioration than ropes made of natural fibers such as cotton or hemp. To secure the rope, attach screw eyes in the corners of the lid and at the front inside edges of the box. Knot the rope to hold it in place at an appropriate length.

Continuous Flow Systems

Continuous flow systems consist of a funnel-shaped bag suspended from a solid structure or a frame (shown above). The bag must be made of a breathable material, wide at the top and narrower at the bottom. Food scraps are added to the top where the worms feed, and the vermicast is harvested at the bottom. These systems are efficient and simple. You can find instructions to make your own via the resources listed on page 176.

CEDAR PATIO BENCH WORM BIN

An adaptation of the plywood bin is this one made out of cedar fence pickets. To make this bin, you will need all of the tools listed for the plywood Patio Bench Worm Bin, with the addition of a table saw. Fence pickets are not dimensional — that is, not all the same width. Because they vary in width, you will need to plan carefully to ensure all four sides are the same height. Materials for a cedar Patio Bench Worm Bin are the same as the one made out of plywood with the exception that you replace the plywood with cedar pickets.

STEP 1. Cut the framing lumber as shown in the Framing Lumber Cutting Diagram on page 25.

STEP 2. Assemble the 2×4 base as explained in step 3 of the plywood patio bench.

STEP 3. Fasten 4' pickets to the base. You will need approximately four pickets; however, because the pickets will not all be the same width, you may need to rip one down to cover the base.

STEP 4. Secure the corner supports to the front and back. As in step 4 for the plywood box, lay the corner supports down on a large, flat surface and secure the pickets to the corner supports. You will need approximately three pickets for both the front and the back.

STEP 5. Cut the side pickets. Again, you will need approximately three pickets for each side.

STEP 6. Assemble the box. You will need a helper to complete the box. Holding up the front, secure each side one at a time to the front, then secure the back to the sides to complete the box.

STEP 7. Drill ventilation holes in the side panels. Follow the same pattern as for the plywood box.

STEP 8. Assemble the lid. Follow the same steps as in step 7 for constructing the lid frame with the 2×4 lumber. Once constructed, cut pickets to cover the width of the lid frame. You will need about ten cut to a length of about 25". The exact length may vary.

STEP 9. Attach the hinges.

STEP 10. Provide lid support with rope.

A BONUS

Most cedar pickets come in lengths of 6 feet. Making a 2 × 4-foot cedar Patio Bench Worm Bin produces a lot of waste. Although this waste makes excellent kindling, I also found that the scraps and a table saw provided a way to make birdhouses.

Indoor and Outdoor Furniture

Some people have built worm bins as pieces of furniture, putting them on rolling casters, staining and finishing boards of finer grades of wood, or using a more attractive (and more expensive) grade of plywood, such as birch. You should be aware of two precautions: use exterior grade plywood, and avoid highly aromatic woods. Since the box will be damp most of the time, you don't want the composite wooden layers of the plywood sheet to separate from each other.

At one time, I believed that all aromatic woods, such as redwood and cedar, might be harmful to worms. But worm workers from redwood country tell me that redwood works just fine for worm bins. Cedar worm bins are available from a number of sources and have been shown to work well, too. That said, I would not recommend using southern or eastern red cedar; these members of the genus *Juniperus* are noted for their fragrance. Instead, use western red cedar (*Thuja plicata*) or northern white cedar (*T. occidentalis*). Both of these woods are used in the manufacturing of fence pickets and can be aromatic when worked.

I made a worm box out of pickets nearly a year ago and the worms are still thriving. Unfortunately, I'm not sure which species of cedar was used, because when I bought the pickets the lumberyard workers could only tell me that the board was not pressure-treated lumber. A good test to be sure you want to use this lumber is to cut the board and then smell it. If the odor is overwhelmingly strong, don't use it. Be aware that *T. plicata* contains plicatic acid, which can cause asthma in some people. It is best if you are going to cut cedar boards to wear a mask and reduce your exposure to the cedar dust.

PLASTIC WORM BINS

Not everyone wants to build their own wooden worm bins. They may not have the tools, the time, or a pickup truck to bring home that big sheet of plywood or the cedar pickets to construct one. Sufficient interest exists in vermicomposting that a number of companies now produce various models of vermicomposting bins. Many people purchase plastic storage containers from discount stores; drill holes in the bottom, sides, and/or tops; and make worm bins that work just fine for them. See the Resources section (page 176) for places that provide instructions for making an indoor plastic worm bin.

The perfect worm bin has yet to be designed. Fortunately, many designs work, provided they are well aerated, well drained, and not subject to extreme temperatures.

Tiered Systems

Available in either circular or square designs, these systems use working trays with mesh bottoms that sit one on top of the other. Taking up little space, these bins are ideal for indoor use. Designed within a completely enclosed tiered system, the bottom catchment tray has a spigot for draining

The Original Vermicomposter

The first home vermicomposting unit available commercially in North America was Al Eggan's Original Vermicomposter produced in Toronto, Canada. This lidded plastic bin had holes in the bottom for drainage and sat on rubber foot supports over a drainage tray. Several small vents in the lid provided some ventilation. Another system is the Worm-A-Way, patented by Mary Appelhof in 1991 and now manufactured by Worm Woman Inc. It is an aerated single bin made of recycled plastic and Mary sold it concurrently with the original edition of *Worms Eat My Garbage* and it is still available for purchase today (see Resources, page 176).

excess water. Although this liquid is sometimes called worm tea, it is more accurately called a leachate (see page 89). The base unit has legs that keep the unit raised up from the floor, providing room for a bucket to collect the liquid from the spigot. The homeowner places moist bedding (see chapter 4 for options) and a starting population of composting worms on the first (bottom) mesh tray. Food waste is added to this tray. When the first tray is full, food waste is added to the second mesh tray. The worms find their way up to the food in this level and begin processing it, depositing their castings in the lower level. With several tiers (anywhere between two and seven) to move up and fresh food as an incentive, worms vacate the bottom level, which eventually contains finely processed worm castings. Castings can then be harvested without disturbing the worms.

Mary Appelhof's original Worm-A-Way
was a simple yet effective system.

CHOOSING THE RIGHT BEDDING

MATERIAL

A major component of your home vermicomposting system is bedding. Worm beddings are multifunctional, since they not only hold moisture but also provide a medium in which the worms can work, as well as a place to bury the food waste.

Worm bedding is usually some form of cellulose, a carbon source that provides energy to the organisms that break it down. Since worms will eventually consume the bedding as well as the food waste, bedding must not be toxic to the worms. The most desirable beddings are light and fluffy, the two conditions necessary for air exchange throughout the depth of the container. This exchange helps control offensive odors by reducing the chances that anaerobic conditions will develop.

BEDDING MATERIALS

Many materials make satisfactory beddings. Mixtures may be used, but a few cautions apply. Some of the more common beddings are listed below, along with brief comments on their advantages and disadvantages. Your choice can be highly individual, depending upon availability, convenience, and economic considerations.

If left six months or more, all the bedding may be converted to vermicast. It can become so dense that the worms have a hard time moving through it. When this material is partially dried and then screened, it is impossible to identify either bedding or the food waste that was originally buried. However, normal procedures for maintaining a healthy worm population (see page 93) require that worms be removed from the bedding while it is still vermicompost, prior to complete conversion to vermicast.

Shredded Newspaper

The least expensive and most readily available bedding is newspaper strips you shred by hand. By fully opening a section of newspaper, tearing it lengthwise down the centerfold, gathering the two halves, tearing it lengthwise again, and repeating the process five or six times for each section, you will get strips ranging from 1" to 3" wide. It doesn't take long to fill your bin with bedding and accumulate enough in a large plastic bag to change the bedding in a few months.

ADVANTAGES

- No cost
- Readily available
- Odorless
- No dust

DISADVANTAGES

- Making strips requires preparation time.
- Inked newsprint can be dirty to handle.
- Large strips dry out more readily than machine-shredded paper.
- Strips tend to mat into layers, making it difficult to bury food waste.

Shredded newspaper is the most readily available bedding.

The most commonly asked question about newsprint is, "Isn't the newspaper ink harmful to the worms?" No, as long as we limit the question to black ink. The basic ingredients of black ink are carbon black and oils, neither toxic to worms. With increasing numbers of newspapers using soy inks, the oils are of even less concern.

Colored inks, however, may be a problem. At one time, heavy metals such as lead, barium, chromium, and cadmium were major components of their pigments. Government regulations and increasing environmental awareness of these substances' toxicity greatly reduced the use of heavy metals in pigments. When I burn a newspaper insert in my fireplace and the flames turn beautiful blue and green colors, they tell me some heavy metals may still be present. Since enough black ink and lightly colored newsprint is available, I try to avoid using heavily colored or glossy advertising sections in my worm bin.

Paper-shredding machines produce high volumes of good worm bedding from newsprint, which are then easily moistened. Paper shredders are clean and easy to use but may be cost-prohibitive for some people.

Leaf Detritus

The bottom of a pile of decaying leaves can yield a satisfactory bedding in the form of partially decomposed leaves. If they are wet, you'll probably even find some worms! Maple leaves are preferable to oak, which take a long time to break down.

ADVANTAGES

- No cost
- Natural worm habitat

DISADVANTAGES

- Unwanted organisms may be present.
- Leaves can mat together, making it difficult to bury food waste.

Animal Manures

Composted horse, rabbit, or cow manures are good bedding for worms. Manure is a natural worm habitat, but it may be difficult to obtain. **Manure should not come from recently dewormed animals** because the drugs used to kill parasitic worms may kill your redworms as well. Some people object to the initial odor, although that should disappear within a few days after redworms are added. Manure is likely to contain other organisms, such as mites, sow bugs, centipedes, or grubs, which some people would rather not have in their homes (see page 113 for dealing with pests in your worm bin).

Setting up a bin with composted manure requires two precautions: First, you need to be sure that the manure has been composted enough that when you add water to it doesn't heat up. If not sufficiently composted when water is added, the micro-organisms will begin the decomposition process again; during that time their body heat increases the temperature of the bedding. Second, to be sure that any composting process doesn't kill your worms, wait a few days before you add worms to the bin.

A forkful of moistened manure can be placed on top of other bedding periodically to help revive a waning culture. Worms will invade the mass and thrive on the nutrients available.

ADVANTAGES

- Can be free for the hauling
- Is a natural worm habitat
- Contains a variety of nutrients
- Makes good castings

DISADVANTAGES

- May not be readily available
- Initial odors possibly objectionable
- May contain unwanted organisms
- May initially heat up, delaying the time when worms can be added
- Can compact easily

Coconut Fiber

Sometimes known as coir, coconut fiber is a clean, easy-to-prepare worm bedding that is becoming more popular as it becomes more available. It comes to market as a block of compressed fiber that expands rapidly when placed in the appropriate volume of water. The moistened fiber is then transferred to the worm bin, where the worms seem to thrive in it. Less acidic than peat moss (pH 5 compared to pH 3.9), coir has a high water-holding capacity and is not supposed to decompose as fast as peat. One woman noted, however, that the worms liked the coir so well they didn't go after the food she was burying in her worm bin! Coir can be effectively mixed with any of the other beddings described to aid water retention or to make manure beddings less dense. I recommend using one-third to one-half coir with other bedding.

Much of the coir is a waste product of the Sri Lankan coconut industry, and large accumulations of this natural fiber present disposal problems on the island. Exporting it as a substitute for peat moss reduces waste disposal problems, provides much-needed income, and helps preserve the limited resource that peat moss represents. Because coir is transported long distances, however, its use in worm bins is not environmentally benign.

ADVANTAGES

- Moisture-retaining
- Clean
- Odorless
- Good for mixing with other beddings

DISADVANTAGES

- Must be purchased
- Transportation may have high environmental impact

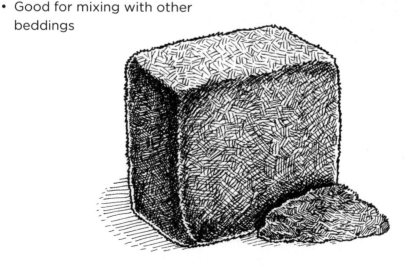

Wood Chips

Worm bin users report that hardwood chips make excellent worm bedding. Mixed with leaves or other bedding materials capable of holding moisture, wood chips provide bulk and create air spaces throughout the bedding. Unlike other beddings, which are consumed by the worms and become vermicompost, wood chips can be reused by screening them when the vermicompost is harvested from the bin.

ADVANTAGES

- Clean
- Odorless
- Good for maintaining aerobic conditions in bin
- Good for mixing with other beddings
- Reusable

DISADVANTAGES

- Quick to dry out
- Availability limited to those with chipper/shredders

Peat Moss

Most of the peat moss used in the United States comes from Canada. There is much controversy regarding the sustainability of harvesting peat moss. Proponents argue that it is a regulated industry in Canada and that only a small portion of the peat bogs are harvested annually. Opponents argue that it is important to understand the ecology of peatlands: They are primarily made up of moss that has decomposed slowly in the absence of air, and the rate of decomposition is less than a millimeter a year. This is the equivalent of 1 meter per 1,000 years. The argument that this slow process to replenish the bog makes peat moss unsustainable is only one part of the reasoning. These bogs are wetlands that filter water and are home to rare native plants. Additionally, as these mosses grow, they absorb carbon. When harvested, the carbon in peat combines with oxygen to produce carbon dioxide, which is released as a greenhouse gas. Since there are other materials available for bedding that do not have these environmental impacts, I do not recommend peat moss for worm bedding.

ADDITIONS TO BEDDING

There are a few materials you can add to the primary bedding to help control moisture, acidity, and texture. Like most other worm-bin strategies, they are dependant upon your circumstances.

Soil

You may have noticed that I have not mentioned using dirt or soil for bedding. In nature, redworms are litter dwellers; that is, they are found among masses of decaying vegetation such as in fallen leaves or manure piles or under rotten logs. Redworms are present in mineral soils only when large amounts of organic materials are also present. Although one investigator recommends using a thin (½") layer of soil in a container holding worms, I have not found that quantity to be essential for home vermicomposting systems. In fact, a big disadvantage of soil is its weight. With just ½" of soil, for example, a container is extremely heavy. I do recommend, however, adding a handful or two of soil when initially preparing the bedding. This provides some grit to aid in breaking down food particles within the worm's gizzard. It also introduces an inoculum of a variety of soil bacteria, protozoa, and fungi, which will aid the composting process.

Calcium Carbonate

Powdered limestone (calcium carbonate) can also be used to provide grit. It has the further advantages of helping to keep conditions in the bin from becoming too acidic and providing calcium, which is necessary for worm reproduction and survival.

You could use the kind that can be mixed with feed or used to line athletic fields, but pulverized eggshells serve the same purpose. Since I add eggshells to my worm bins regularly, I frequently don't bother with the limestone.

CAUTION: Do not use slaked or hydrated lime. The wrong kind of lime will injure your worms and may kill them.

Rock Dust

I highly recommend rock dust as another source of grit. Also called rock powder or rock flour, this is finely ground rock from natural or industrial processes. Natural processes include the grinding of glaciers against rock surfaces, forming sand and silt as a consequence of erosion. Industrial processes would be gravel-crushing operations to produce aggregate for the construction industry, resulting in a powdery by-product called fines. Depending upon the source rock, rock dust can contain many trace minerals that support plant growth. Combined with the action of microorganisms in a worm bin, the availability of minerals in rock dust enhances plant growth more effectively than if the rock dust were not present in the vermicompost at all.

Zeolite

Zeolite is a mineral used in granular or powder form in commercial worm beds in Australia and New Zealand to balance pH and absorb ammonia and other odors. Of volcanic origin, zeolite's natural negative charge attracts positively charged odor molecules and absorbs them on its surface. Using zeolite in a worm bin would provide grit for the worms' gizzards in addition to these other benefits.

USING THE RIGHT KIND OF WORMS

Most people think that "a worm is a worm is a worm." In fact, there are many kinds of worms, each with different jobs to do. It is important to use the right worms in your home vermicomposting system.

Most of the worms that you could dig up from your garden would not be suitable for vermicomposting. You want a worm that processes large amounts of organic material. The worms should reproduce quickly in confinement and tolerate the disturbance caused when you lift the lid to bury food waste or add bedding. When small organisms are raised in a controlled environment, they are said to be cultured; the culture of earthworms is known as vermiculture.

TYPES OF WORMS

The soil-dwelling species of worm, or "earthworkers," don't process large amounts of organic material like the "composters." They don't reproduce well in confinement, and they won't thrive in a worm bin if you dig around and mess up their burrow system. Understanding the characteristics of different worms and why common names can be confusing is the purpose of this chapter.

Redworms are the most satisfactory kind to use in your home vermicomposting system, but what I call "redworm" you may know as "red wiggler." Your neighbor may call it a manure worm. The bait dealer down the road may refer to it as a red hybrid. Other common names for this same animal are fish worm, dung worm, fecal worm, English red worm, striped worm, stink worm, brandling, and apple pomace worm. A distinct pattern of alternating red and buff stripes characterizes some of these worms, hence another common name, tiger worm. Calling earthworms by common names can cause communication problems. With so many names, how can any of us know when we are talking about the same worm?

All earthworms belong to one of three groups, depending on their behavior in their natural environment.

EPIGEIC (Greek for "upon the earth") worms are commonly called composters. They live and feed in surface mulch areas on decaying matter. These worms are usually small, reproduce rapidly, and, because they have pigmentation, are reddish brown in color. They do not burrow and are the type of worm used in vermicomposting. Some scientists further divide this group into those worms that live mostly in compost, such as *Eisenia fetida,* and those that live on the surface of soil in leaf litter, such as *Lumbricus rubellus.*

ENDOGEIC (Greek for "within the earth") worms live in the topsoil. Ranging in size from small to large, these worms lack pigmentation. They will appear to be blue-gray, yellow, pink, or a whitish color. These worms feed on soil and decaying organic matter and form a network of horizontal burrows.

ANECIC (Greek for "out of the earth") worms live in the subsoil, making vertical burrows as deep as 6 feet. These large worms feed on fresh surface litter that they drag down into their burrows. They deposit their waste (castings) at the surface. Like epigeic species, they also have pigmentation, causing them to be reddish brown in color.

What's in a Name?

A marketing strategy for some worm growers is to create a name that establishes their worms as unique or better than others so the growers can justify a higher price. Such names as Jumbo Worm or Super Giant come to mind. These are probably not different species, and certainly not hybrids (combinations of species), but are simply likely to be well-fed worms. If you are just buying some worms for fishing, it probably doesn't matter. Superworm is the common name for the larvae of the darkling beetle and is sold as pet food for reptiles. *Zophobas morio* is the scientific name.

But if you are going to use the worms you buy to set up a worm composting system, make sure the grower can give you the scientific name. That way you can be sure that the worm you purchase is suitable for composting and not an invasive. And if you need to learn more about it, you can read some of the scientific papers about your worm of choice.

Scientific Names

To be certain they are talking about the same thing, scientists have developed a precise system for naming organisms. Since much information in this book comes from scientific papers, I will be using scientific names. So that you won't be confused when I do, here are some basic rules that all scientists follow:

- The name of each organism consists of two words, the first of which is called a genus (plural: genera); the second, the species. All organisms of the same genus are more closely related than those of different genera. For instance, human beings are members of the species *Homo sapiens*.

- Correct usage requires that the genus name always be capitalized and the species name be lowercased.

- Both terms are either Latin or Greek in origin. Sometimes they are Latinized versions of words in other languages. Genera and species are italicized in print and underlined when italics are not available.

The International Commission on Zoological Nomenclature provides and regulates a uniform system of zoological nomenclature (a system of names), ensuring that every animal has a unique and universally accepted scientific name. As more research is done on a species, it is possible that not only the species name can change but in some cases the genus can change. An excellent website to research earthworm names is DriloBASE Taxo. The contributors include scientists who are well known in the vermicomposting community, and the site lists synonyms to clarify the names of the worms being researched (see Resources, page 176).

COMPOSTING WORMS

Certain epigeic worms can eat and digest organic material quickly. Although they have a short life cycle, their reproductive rate is high. Additionally, they tolerate handling and a wide range of environmental factors. Although there are thousands of species of worms, only a few have been used extensively in vermicomposting. Let's examine those now.

TEMPERATE SPECIES

Redworm
Eisenia fetida

The worms I use are *Eisenia fetida* (which I pronounce as "i-SEE-nee-uh FET-id-uh"), said to give off a fetid odor when roughly handled. The other common names for this worm are brandling or tiger worm. It has a red-buff transverse segmented stripe. These worms process large amounts of organic material in their

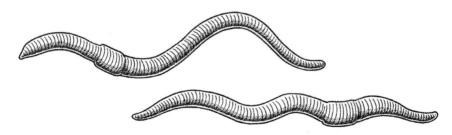

ANECIC

ENDOGEIC

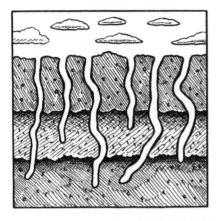 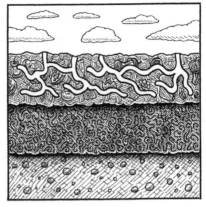

Endogeic worms live and work on the surface in upper organic layers of topsoil. The soil-dwelling anecic worms move from one layer to another, mixing layers as they go.

natural habitats of manure, compost piles, or decaying leaves. They are fast reproducers and tolerate a wide range of temperatures, acidity, and moisture conditions. They are tough worms and withstand handling well. Because sufficient markets exist encouraging people to culture *E. fetida* on a part- or full-time basis, anyone may purchase them almost any season of the year. They can readily be shipped via package delivery services or through the mail. Many people who vermicompost would like to transfer extra worms from their bins to their gardens to increase the worm populations there, but I never recommend this for *E. fetida* because they are not soil dwellers.

Red Tiger
Eisenia andrei

Eisenia andrei is a relative that lacks the red and buff striping of the redworm. Known as the red tiger, *E. andrei* (abbreviating the genus name to just its initial saves space once it has already been spelled out in full) has slightly better performance characteristics then *E. fetida*. Many times this species is incorrectly called *E. fetida*. It has been determined that *E. fetida* and *E. andrei* are two different species; however, they are easily confused with one another. Although DNA barcoding tests can prove that *E. fetida* is

E. fetida, the same test will sometimes identify *E. andrei* as *E. fetida*. Most commercial cultures contain a mixture of both species, and growers do not separate them.

TROPICAL SPECIES

Indian Blue Worm
Perionyx excavatus

Perionyx excavatus is suitable for vermicomposting in warm climates. Called the Indian blue worm in some regions, it is a tropical species that reproduces well in culture and tolerates handling. Scientific investigations show that *P. excavatus* is intolerant of cold, so it would not live outside through the winter in a cold climate. It also has a tendency to move out of worm bins for no apparent reason, a characteristic my staff acknowledges by calling them "travelers."

African Nightcrawler
Eudrilus eugeniae

Another restless worm used for vermicomposting, *Eudrilus eugeniae* is large and commonly known as the African nightcrawler. As its common name suggests, this species has a tropical origin. The worms can reproduce quickly and process large amounts of organic material rapidly within their optimal temperature range of 59 to 77°F (15 to 25°C); temperatures below 50°F (10°C) kill them. *E. eugeniae* is therefore limited to warm climates or heated buildings.

INVASIVE WORMS

Invasive species are those that are nonnative to an ecosystem and whose introduction causes or is likely to cause economic or environmental harm or harm to human health.

Eleven thousand years ago, northern North America was covered in ice. This glaciation destroyed native worms. The forests in that part of the world evolved without earthworms. With the arrival of European settlers in the 1800s, earthworms began to enter North America again. Since that time, they have continued to be distributed and spread through soil redistribution, in plant pots, after being used as bait, and via mud on vehicles and boots. On their own, worms move slowly. Their rate of movement has been calculated at less than 30 feet per year. It would take an earthworm over a thousand years to move just over 5 miles! It is people who brought earthworms to northern North America.

About three decades ago, scientists in Minnesota and New York noticed that the forest floors were changing. Since then, studies have shown that some worms can have a negative impact on ground-nesting birds and forest regeneration and increase the spread of invasive plants such as garlic mustard.

Asian Jumping Worm
Genus *Amynthus*

Jumping worms are sold for bait and as a composting worm. Common names include Alabama, Georgia, Jersey, or Asian jumpers. Some call this a snake worm or crazy worm because they move like a snake when picked up. This worm can eat and process more than its body weight in organic material each day. They tend to process leaves and soil until the soil becomes dry and granular. Originating from Southeast Asia, the worms are a problem in many parts of North America. These worms were found in 75 percent of the test plots studied in the Great Smoky Mountains National Park. The park now has a regulation that no live bait can be used for fishing within its boundaries. Earlier in 2009, all 51 species of this genus were listed by the state of

Wisconsin as prohibited species, a designation made to keep the worms from coming into the state. When the Wisconsin Department of Natural Resources discovered jumping worms in the state in 2013, their designation was reclassified as "restricted species." Neither prohibited nor restricted species may be sold, introduced, transported, or propagated in Wisconsin.

Red Earthworm
Lumbricus rubellus

At one time, *Lumbricus rubellus* was considered suitable for worm composting. Some consider *L. rubellus* to be the "true" redworm; others call it the dung worm or the red marsh worm. It has been found in compost heaps and manure piles, and in pastures, particularly under cow patties or manure. Some worm growers claim to have *L. rubellus* in their beds, but scientists consistently report that every time they have checked those claims, the worms have been *E. fetida*, *E. andrei,* or a mixture of both species. Studies regarding the maturation and reproductive rate of this worm have shown it to be less than ideal as a vermicomposter. *L. rubellus* looks similar to a nightcrawler but is smaller. The top anterior of the worm is purple-maroon, and it is lighter at the tail and underneath. It tends to be thicker-bodied between the clitellum (the region that secretes cocoon material) and the head than other worms. It is commonly sold as fish bait. Because it feeds in the litter layer and in the top few inches of soil, it is sometimes referred to as an epi-endogeic species. Great Lakes Worm Watch lists this worm as problematic in their forest, as it can rapidly degrade and remove the forest floor.

How to Prevent the Spread of Invasive Worms

In 1881, Charles Darwin published *The Formation of Vegetable Mould through the Action of Worms, with Observations on their Habits*. This book showed us how remarkable earthworms are. They are amazing creatures that can increase soil fertility and aerate soil with their movement. Those of us who are worm workers know and appreciate what benefits earthworms provide, but even a good organism can become an invasive when it is not handled with care.

What do we as worm workers do to stop an invasive species? First of all, I only recommend *Eisenia fetida* or *E. andrei* for home composting. Since these worms feed on the top layer of the soil, they do not survive winters. Second, I recommend following the directions of national and state agencies regarding these worms. These directions usually include the following:

- When fishing, arrive clean and leave clean, including your boots. Pay special attention to tire threads that can hold soil.

- Learn to identify different species of worms. There are many excellent online resources (see page 176).

- Educate yourself on invasive species.

- Pay attention to your landscape and gardening materials to be sure they are free of any problematic organisms.

- Never discard unused bait by dumping it outside.

- Use noninvasive bait.

- Freeze the vermicompost before you use it outside the bin.

OTHER WORMS IN NORTH AMERICA

If you dig up your garden, you will mostly likely find species of worms other than those described in this chapter (aside from *Lumbricus terrestris*, below). In the northern United States and southern parts of Canada, about 90 percent of the worms would be one of eight soil-dwelling species, including *Aporrectodea caliginosa*, *A. trapezoides*, *A. tuberculata*, *Dendrobaena octaedra*, *Dendrodrilus rubidus*, *L. terrestris*, *L. rubellus*, and *Octolasion tyrtaeum*. More species will be found in the southern part of the United States, which was not covered by glaciers 11,000 years ago.

Identifying these species would require suitable magnification, a good pair of forceps, and a tool for pointing or probing. Depending on the guide or identification key you use, preliminary determinations of species can be made on the basis of pigmentation, general body shape, length of the worm, and position of the clitellum. You may have to identify the type of projection over the mouth, locate the position of various openings for sexual organs, and determine the pattern for setae (bristles) on each segment. We need more people who can identify kinds of worms. I encourage you to learn how to do so.

Canadian Nightcrawler
Lumbricus terrestris

Today more than 5,000 species of earthworms have been classified, and of those, *Lumbricus terrestris* was the first. Commonly called the Canadian nightcrawler, these worms are widespread in Europe and North America and are found in parts of New Zealand and Australia. Most people recognize the nightcrawler, and it has been called many names, including the dew worm, night walker, rain worm, angle worm, orchard worm, and night lion.

Nightcrawlers are not suitable worms for the type of home vermicomposting system described in this book. I once placed 80 nightcrawlers in my worm bin along with the redworms already there. Two months later, I found only one live night-crawler, and it was immature. Although satisfactory environments can be created for nightcrawlers indoors, they require large amounts of soil, and the bed temperature cannot exceed 50°F (10°C). For obvious reasons, those conditions are very impractical for a home vermicompost bin. Nightcrawlers dig burrows and don't like to have their burrows disturbed. If you try to bury food waste, nightcrawlers move quickly around the surface of the box trying to escape your digging.

Nightcrawlers do, however, play important roles in soil fertility in some ecosystems. These large soil-dwelling earthworms have extensive burrows extending from the ground surface to several feet deep. They come to the surface on moist spring and fall nights and forage for food, drawing dead leaves, grass, and other organic material into their burrows, where they feed upon it at a later time. Nightcrawlers perform important soil-mixing functions. They take organic material into the deeper layers of the soil, mix it with subsoils that they consume in their burrowing activities, and bring mineral subsoils to the surface when they deposit their casts. Through their burrows, nightcrawlers also aid in soil aeration and in water retention by increasing the rate at which water can penetrate the deeper soil layers. They may not be good for your worm boxes, but they are very good for your gardens. They are, however, not good for the northern deciduous forests, as they degrade the forest floor and can enable the spread of invasive plants in some regions.

$ ACQUIRING YOUR WORMS

Before determining the number of redworms you need to start vermi-composting, and where to get them, an understanding of their amazing reproductive potential is helpful.

One reason for using redworms is that they reproduce quickly. When some people learn how rapidly they reproduce, they become concerned that redworms will overrun their bin, but the worms' numbers are controlled by environmental factors.

HOW DOES A WORM REPRODUCE?

Earthworms have both male and female organs, yet most still need to mate in order to reproduce. To understand why mating is even necessary when each worm produces both eggs and sperm, it is helpful to first understand their physiology. Before explaining how this works, know that self-fertilization has been reported in *Eisenia fetida*, and that worms in the genus *Amynthus* always self-fertilize.

Mating

The swollen region about one-third of the distance between the head and tail of a worm is the clitellum, sometimes known as the girdle, band, or saddle. The presence of a clitellum indicates that a worm is sexually mature. Bait worms with this structure are commonly called "banded breeders," showing that they are old enough to breed and produce offspring. Just as worm species differ in external characteristics, they differ somewhat in mating behavior. For example, nightcrawlers extend themselves from their burrows to seek another nightcrawler with which to mate. Attracted by glandular secretions, they find each other and lie with their heads in opposite directions, their bodies closely joined. Their clitella secrete large quantities of mucus that forms a tube around each worm. Sperm from each worm move down a groove into receiving pouches of the other worm. The sperm, in a seminal fluid, enter the opening of sperm storage sacs, where they are held for some time.

Redworms differ from nightcrawlers by mating at different levels in their bedding rather than just upon the surface. Under proper conditions, they can also be observed mating at any time of year, whereas some species mate only during particular seasons.

Sometime after the worms separate, the clitellum secretes a second substance, a material containing albumin. The

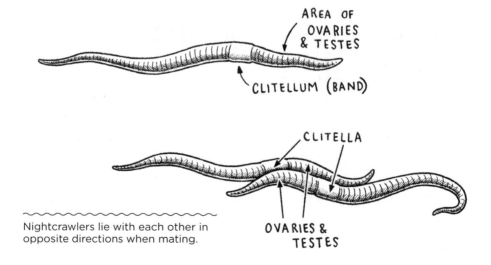

AREA OF
OVARIES
& TESTES

CLITELLUM (BAND)

CLITELLA

OVARIES &
TESTES

Nightcrawlers lie with each other in opposite directions when mating.

albuminous material hardens on the outside to form a cocoon in which eggs are fertilized and from which baby worms hatch. As the adult worm backs out of this hardening band, it deposits eggs from its own body and the stored sperm from its mate, which can fertilize eggs for several cocoons. Sperm fertilize the eggs inside this structure, which closes off at each end as it passes over the first segment. Sometimes called an egg case or capsule, this home for developing worms is more properly called a cocoon.

Cocoons and Baby Worms

About the size of a matchhead or a small grain of rice, cocoons are lemon-shaped objects. They change color as the baby worms develop, starting as a luminescent pearly white, becoming quite yellow, then light brown. When the hatchlings are nearly ready to emerge, cocoons are reddish. By observing carefully with a good hand lens, it is sometimes possible to see not only a baby worm but the pumping of its bright red blood vessel. The blood of a worm is amazingly similar to ours, having the same iron-rich hemoglobin as its base to carry oxygen.

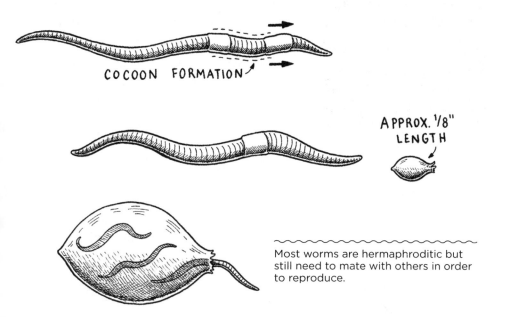

COCOON FORMATION

APPROX. 1/8" LENGTH

Most worms are hermaphroditic but still need to mate with others in order to reproduce.

Best Guesses

How quickly will your worm population grow? Dr. Roy Hartenstein of Syracuse, New York, has calculated that eight individuals could produce about 1,500 offspring within six months' time. He based this upon their producing two cocoons per worm per week, of which 82 percent hatch and average 1.5 hatchlings per cocoon. If they reach maturity at 5 to 6 weeks and continue producing cocoons for 40 to 50 weeks of fertility at 77°F (25°C), his calculated total would result. Calculations based upon other assumptions would result in different projections.

It takes at least three weeks' development in the cocoon before one to several baby worms hatch. The time to hatching is highly dependent on temperature and other conditions. I worked with worms for years before I ever saw a worm emerge from its cocoon. I have observed hatchlings work their way out of their cocoon, thrashing about vigorously. When I turned on bright lights to try to photograph them, they quickly retreated, reacting negatively to light just as adult worms do. I have since been able to videotape baby worms twisting and turning inside their cocoons. Imagine how excited I was to capture an hour-long sequence of four baby worms hatching from their cocoon!

Newly emerged worms are whitish and nearly transparent, although the blood vessel causes a pink tinge. They may be from 1/2" to nearly 1" long when they hatch, but they weigh only 2 to 3 mg. At that size, it would take more than 150,000 hatchlings to make 1 pound (0.5 kg) of worms.

Although each cocoon may contain as many as ten fertilized eggs, normally only two or three hatchlings emerge. The number of hatchlings varies depending on such factors as the age of the breeder that deposited the cocoon, its nutritional state, the temperature, and whether the temperature is constant or fluctuates daily. This knowledge makes it possible to establish conditions for greater hatchling production.

Reaching Maturity

The time it takes for a baby worm to become a breeder also varies, depending on temperature, moisture, food availability, and population density. A redworm can be sexually mature and produce cocoons in eight weeks, but ten is more common. Once it breeds and begins laying cocoons, it can deposit two or three cocoons per week for six months to a year. Conservatively, then, if a two-month-old breeder laid two cocoons a week for 24 weeks, and two hatchlings emerged from each cocoon, one breeder would produce 96 baby worms in six months (2 cocoons × 24 weeks × 2 hatchlings).

The situation is more complicated than that, however. Before the first two months are up, the first hatchlings will be able to breed. These could produce two cocoons for 16 weeks with two hatchlings coming from each of the four worms resulting from the original breeder's first-week's production. Some of the 256 more worms produced during this time will die before the six-month period is up. The math quickly gets complicated, but since optimal conditions for such geometric increases in numbers will never be achieved, theoretical projections are more confusing than informative.

Population Controls

With the reproductive potential described, we come back to the question of why worms don't take over the bin. Three basic conditions control the size of a worm population: availability of food, space requirements, and fouling of their environment.

When food waste is fed regularly to worms in a limited space, the worms and associated organisms — both microscopic and larger — break down this waste. They use what they can and excrete the rest. As the worms reproduce, the voracious young worms compete with their parents and all the other worms in the culture for the limited food available. Additionally, all the worms excrete wastes in their castings, which have been shown to be toxic to members of their own species. As time goes on, more worms compete for the limited food, and more and more of the bedding becomes converted to castings. The density of the worms may exceed the number favorable for cocoon production, and reproduction slows down. Undesirable conditions in their continually changing environment may cause some worms to die. Some worms will die of old age. Interestingly, you will rarely see dead worms, because they are rapidly decomposed by other associated creatures in this active composting environment.

The controls you exert over your worm population will affect this whole process. You may choose to feed an ever-increasing population more and more food. If you want more and more worms, you will eventually have to provide them more space and fresh bedding and enable them to get away from high concentrations of their castings. You may even choose to become a worm grower and try to keep up with the ever-increasing demands for food, space, and timely elimination of accumulated end products. But that's another story, and it's more complicated than simply keeping enough worms alive to process your kitchen waste so that you can use the rich end product to grow healthier vegetables and house plants.

DECIDING HOW MANY WORMS YOU NEED

Quantities of worms are specified here in terms of weight rather than numbers for two main reasons, one ancedotal and the other biological. The anecdotal reason is that the first season I sold worms, my partner and I sorted and counted 50,000 worms — one by one by one. If you ever have to count 50,000 of anything, you'll find an easier way to do it, too. From that time on, I have sold worms by weight. Of course, the first question I always get is, "How many worms are in a pound?" Although the number will vary depending on the size of the worms, there are some guidelines.

Worm growers commonly estimate that there are about 1,000 breeders per pound (2,200/kg) for young redworms. A bait dealer selling worms for fishing would prefer that the worms be considerably larger than that. If redworms don't run between 600 to 700 per pound (1,300 to 1,500/kg), customers complain that they are too small to get on a hook.

Biologically, it is the weight of worms, not their number, that is important in vermicomposting. Worms can consume more than their weight each day, regardless of their size. During their rapid growth stage, juveniles eat proportionally more than adults. Think of feeding a teenager in your home. Since many small worms can move as much material through their intestines as fewer large worms of equal mass, what you want is an earthworm biomass sufficient to do the job.

With some idea of how fast redworms reproduce, you might conclude that your worm bin would eventually produce enough worms to handle all your food waste regardless of how many worms you used initially. You could start with a few dozen redworms and regularly feed them small amounts of food waste. If patience is one of your virtues, you could wait until their natural tendency to reproduce under proper conditions yields several thousand worms. However, most of you will want your worm composting system to handle enough of your organic kitchen waste from the day you set it up to assure you that it really works. As the worm population increases, you will be able to add more food waste.

Three factors influence the number of worms you will want to start with:

1. The average amount of food waste to be buried per day

2. The size of the bin

3. The cost of worms and the amount of money you want to spend

Since your bin size will be based upon how much food waste you expect, the amount you will be burying is the critical factor. I usually recommend starting with at least a pound (0.5 kg) of worms.

Worm-to-Daily-Waste Ratio

The relationship between weight of worms required and a given amount of food waste can be expressed as a "worm:waste ratio" (the colon here stands for the word *to*). I recommend a worm:waste ratio of 2:1, based on the initial weight of worms and the average daily amount of waste to be buried. Thus, if you generate 7 pounds (3.2 kg) of food waste per week, you would average 1 pound (0.5 kg) of food waste per day. You will eventually want about 1 pound of worms per cubic foot of volume in your bin. Since they will reproduce, you can start with about half that quantity. For 7 pounds of food waste per week, I would start with 2 to 3 pounds (0.9 kg to 1.4 kg) of worms in a 6-cubic-foot worm bin.

Calculations for a household that produces a smaller amount of food waste are similar: 3.5 pounds waste/week = 0.5 pound waste/day on average (1.6 kg, 0.2 kg).

Following the suggested worm:waste ratio of 2:1, use 1 pound of worms for 0.5 pound of daily food waste (0.5 kg worms for 0.3 kg daily food waste). As explained in chapter 3, your container size should provide a surface area of about 4 square feet, or 1 square foot for each pound of food waste per week. I would set up a 2×2-foot worm bin with 1 pound of worms. Four square feet would require 4 pounds (1.8 kg) of food waste. Using the 2:1 ratio, this would be 2 pounds (0.9 kg) of worms — but because they reproduce so quickly, you can use half that amount, or 1 pound (0.5 kg).

SOURCES OF REDWORMS

Most people order their worms online. Be sure and check out multiple websites. See if you agree with what they say on their site. Do they know the scientific name? If two companies are equal in knowledge and reputation, use the one closest to you to save on shipping time. You may be able to find garden centers or local growers to provide your initial stock. **If you buy redworms from bait dealers, expect to pay about twice as much as you would if you buy directly from a grower.**

Another way to obtain the redworms you need to set up your home vermicomposting system is to order them from one of the commercial earthworm growers who advertise in the classified sections of gardening and fishing magazines. Redworms are easily packaged and shipped through the mail or by package delivery services. Some growers advertise and ship year-round, others seasonally. As with growing conditions, temperature extremes should be avoided, but if the temperature is colder than 20°F (−6°C) or above 90°F (32°C), growers will usually wait until these temperatures moderate before they ship the worms.

Bed-Run Worms

Many redworm growers sell what they refer to as *bed-run*, also known as pit-run or run-of-pit worms. These are worms of all sizes and ages, from which bait-size worms may or may not have been removed. Since there could be between 150,000 to 200,000 hatchlings in a pound (330,000 to 440,000/kg), the number of bed-run worms in a pound will vary tremendously; however, 2,000 is a figure commonly used (4,400/kg).

There are no hard-and-fast rules to tell you whether to start with breeders or bed run. Breeders will lay cocoons more quickly and increase the number of individuals sooner, but they usually cost more from commercial growers because of the labor required to sort them. Also, some growers think breeders take longer to adjust to new culture conditions than do bed-run worms.

Bed-run worms are a mix of juveniles and adults and can be shipped through the mail in their own bedding.

Free-Range Worms

Those of you who like adventure may be able to collect redworms from a compost pile or from their natural habitat. Your chances increase if you have a friend with horses, a barn, and a manure pile. You may have to turn over a lot of manure to find any, but then again, you could get lucky and find hundreds of worms in a few pitchforks full of moist manure at just the right stage of decomposition to support them.

If you can order bed run by weight, you will certainly get more worms than if you purchase breeders by weight. These young, small worms will grow rapidly and be able to reproduce soon. If they adjust to their new home faster than breeders would have, you will be ahead starting with bed run, especially if they are cheaper.

Whichever you start with, breeders or bed run, when they produce more worms than the food waste you are feeding them will support, many will get smaller, some will slow reproduction, and others will die. Eventually, no matter how many worms you start with, the population will stabilize at about the biomass that can be supported by the amount of food they receive.

When you have completed tasks 1 through 6 on the checklist, you are ready to set up your worm bin.

You have determined approximately how many pounds of kitchen waste you dispose of per week, purchased or built your container accordingly, selected and obtained your bedding, and ordered or collected your worms. If all materials are on hand, it takes about an hour to set up your bin.

- ☑ Read *Worms Eat My Garbage*.

- ☑ Weigh organic kitchen waste for two to three weeks to get the average amount produced in your household so that you can determine the right size bin for the amount of waste you produce.

- ☑ Determine the quantity of worms you need and order worms.

- ☑ Purchase a bin or select the size of container required and assemble the materials.

- ☑ Determine what beddings are available, and either order materials or scrounge.

- ☑ Build or assemble your bin.

PREPARATION OF WORM BEDDING

NEEDED

- Completed worm bin
- Bedding materials (see chapter 4)
- 1 or 2 handfuls of soil
- Bathroom or utility scale
- Jug
- Large clean plastic or metal garbage can for mixing bedding

The amount of bedding you need depends, of course, on the volume of your container. This can be a very rough measure. It is important to prepare enough bedding initially so that your container will be about three-quarters full with the moistened bedding in place. The following chart provides the approximate weight of newspaper to set up several of the bins described in this book.

A helpful rule is to use about 3 pounds of newspaper per cubic foot volume of the bin (50g newspaper/liter). For plastic containers that give capacity in gallons, use about $2/5$ pound of paper per gallon. If you don't have a household utility scale, stand on a bathroom scale, first alone and then with your plastic bag full of dry bedding. The difference between the two weights, of course, is the weight of the bedding.

The major task remaining to set up your worm bin is to prepare the bedding for the worms by adding the proper amount of moisture; bedding should be damp, but not soggy. A worm's body consists of approximately 75 to 90 percent water, and its surface must be moist in order for the worm to "breathe." By preparing bedding with approximately the same moisture content (75 percent) as the worm's body, the worm doesn't have to combat an environment that is either too dry or too moist.

Bed Capacity

Type of Bin	Cubic Feet	Volume in Gallons	Pounds of Bedding
2'×2'×8" box	2.7	20.2	8.1
1'×2'×3' box	6.0	44.9	18.0
Patio Bench	8.6	64.3	25.8
Worm-a-Way	2.3	17.2	6.9

When using shredded paper bedding, a 75 percent moisture content can be easily obtained since the residual moisture present is minimal. Just weigh the bedding and add water equal to three times its weight. To get 75 percent moisture, for example, add 15 pounds of water to 5 pounds of shredded newspaper bedding (6.8 kg water to 2.3 kg bedding). Or, expressed another way:

water:bedding ratio = 3:1 by weight

One gallon of water weighs 8.34 pounds (3.78 kg). In the example above, for 5 pounds (2.3 kg) of bedding you would use about 1.8 gallons (6.8 l) of water. The key to the correct ratio of water to bedding is that, when squeezed, the bedding is damp. The level of dampness is similar to clothes coming out of the washing machine. Coir needs to be soaked first because it can hold 8 to 10 times its weight in water. Moisture in leaf detritus will vary, making it necessary to do the squeeze test to be sure that it is damp enough but not too damp.

Because plastic bins tend to accumulate water, I usually recommend that you use about one-third less water when setting up a plastic bin. I will discuss this more in chapter 9.

Place about half of the bedding into the large mixing container. Add about half of the required amount of water and mix it into the bedding. Then add one or two handfuls of soil and the remaining bedding and water. Mix again until the water is well distributed throughout the bedding. Now dump the entire contents of the container into your worm bin and distribute it evenly. (The bedding absorbs the water so that little, if any, leaks from the holes in the bottom of the bin.) Your bin is now ready for the worms!

MANURE BEDDING

If you are using manure for bedding, it is more difficult to determine how much water to add to obtain the proper moisture content, since you don't know how much moisture is already in the manure. Basically, you want the manure damp, but not soggy.

If you squeeze a handful and produce 3 or 4 drops of water, it's probably all right; 20 drops or a stream of water is too wet.

With manure bedding, remember to add water at least two days before you add worms. Then, if the manure heats up as it begins to compost, the worms won't die from the heat.

GETTING WORMS IN THE MAIL

Most growers package worms for shipping in peat moss, inside a permeable bag inside an aerated box. Experienced shippers pack worms in a fairly dry bedding for two good business reasons:

• Shipping costs are great.

• There is no point in paying to ship excess amounts of water.

It is more important, however, to provide a satisfactory environment for the worms. Although worms need bedding with some moisture in it, too much moisture can intensify the effects of temperature extremes during shipping. In midsummer when the temperature is likely to be 80 to 90°F (27 to 32°C), a drier bedding acts as insulation, plus provides sufficient oxygen for the worms. Too much moisture fills air spaces, and the additional heat stimulates natural microorganisms associated with the worms to use up all available oxygen before the worms can get it. If they die, neither you nor I will want to open the box for the smell!

The insulation effect of a drier bedding for packaging also pertains to cold-weather shipments of worms. Although the worms will lose some of their moisture to the bedding, they are better off than if they were to freeze because it was too moist and too cold.

If you receive worms that seem dry, assume that the worms will quickly regain their lost body moisture when they are placed in a properly prepared bedding. This should be done within a day or two. Responsible growers try to do what's best for the worms, guarantee their shipments, and provide information so that the customer knows what to expect.

If you need to hold your worms more than two days, open the box, sprinkle water on top to make the worms more comfortable, and add a light layer (a tablespoon [15 mL] or so per thousand worms) of oat bran on top. Feed again only when most of the food disappears in one or two days. Don't stir the grain into the bedding, or the bedding may become sour and/or overheat.

Introducing the Worms and Food Waste

When your bedding is ready to receive the worms, open their container and dump the entire contents on top of your freshly prepared bedding. Gently spread any clumps of worms around the surface. Leave the room lights on for a while. The worms will gradually move down into the bedding as they try to avoid the light. Within a few minutes, the majority of worms will have disappeared into the bedding. If any remain on the surface after an hour, assume that they are either dead or unhealthy. Remove them.

Once the worms are down, you may start burying food waste. Of course, by now you know the average amount of food waste your household produces in a week. Dig a hole big enough to accept the amount of food waste you are burying and dump the food waste into the hole. Draw enough bedding over the food waste to cover it completely.

An alternative method is to wrap peelings and other food waste in newspaper to keep them "contained," and add them as a package. This keeps waste tidy and covered and provides additional bedding. This method does require a bit more water.

With lidded containers, merely close the lid after the worms go down. The simpler boxes without lids require a piece of carpet, burlap, or sheet of black plastic to keep out the light and retain moisture. The worms work up to the surface; when you lift the plastic, you will see them scramble down into the bedding.

If your worm bin does not have a lid, you can use black plastic sheeting to keep out light and retain moisture.

What's waste to me or you may be slop for the pigs or food for the dog to someone else. I have previously used such terms as *organic kitchen waste* and *table scraps*, but now it's time to be more specific about what waste you can expect to feed to your worms.

KITCHEN WASTE FROM MEAL PREPARATION

Any vegetable waste that you generate during food preparation can be used: potato peels, grapefruit and orange rinds (see page 77 about using too much citrus), outer leaves of lettuce and cabbage, celery ends, and so forth. Plate scrapings might include macaroni, spaghetti, gravy, vegetables, or potatoes. Spoiled food from the refrigerator, such as baked beans, moldy cottage cheese, and leftover casserole also can go into the worm bin. Coffee grounds are very good in a worm bin, enhancing the texture of the final vermicompost. Tea leaves, even tea bags and coffee filters, are suitable.

Eggshells can go in as they are. I have found as many as 50 worms curled up in one eggshell. Usually, I dry the shells separately, then pulverize them with a rolling pin so they don't look quite so obvious when I finally spread the vermicompost in my garden. Grinding up eggshells also increases their surface area. This makes calcium carbonate more readily available to the microorganisms and other decomposers in the bin and, later, to plants in the garden.

The list below shows some of the variety of food waste that can be fed to worms. It was developed from waste actually buried in worm bins I helped establish at a Michigan nature center during a publicly funded project in the 1970s. Coffee grounds don't appear on the list merely because none of the six participants' families drank coffee. Use this list as a guideline only; it is not, by any means, comprehensive.

FOOD WASTE FOR WORMS

- Apples
- Baked beans
- Banana peels
- Bread
- Cabbage
- Cake
- Celery
- Cereal
- Cheese
- Cream cheese
- Cream of wheat
- Cucumber
- Deviled eggs
- Eggshells
- Grapefruit peels

- Grits
- Lemon peels
- Lettuce
- Molasses
- Oatmeal
- Onion skins
- Orange peels
- Pancakes

- Pears
- Pineapple
- Pizza crust
- Potatoes
- Tea leaves
- Tomatoes

Your worms can eat a hearty mix of food waste.

Citrus Warning

Excess quantities of citrus will kill worms. If your bin is small and you squeeze a dozen or so oranges for juice, I advise you not to put all the rinds in the bin at once. Lemons, oranges, and limes contain limonene, which is toxic to worms. I have occasionally put an orange peel in my bin, and it disappears, but I generally put citrus rinds in my outdoor compost pile.

A Word about Meat Waste and Bones

You will not find any meat on the list of food waste on page 76. When setting up the project at the nature center, we deliberately excluded the burial of meat, in order to avoid foul odors. Because worms do not have teeth, and their digestive enzymes are limited, they rely on the aerobic activity of microbes to start the decomposition of food. This rotting process tends to smell more with meat than with vegetables. We also wanted to avoid attracting insects and rodents (and sometimes even larger scavengers, when the bins were outdoors), prevent possible injury from sharp bones, and enhance the appearance of the final vermicompost. Since the demonstration bins were to be located in a public exhibit area and seen by thousands of visitors, it was important to avoid such potential problems.

Nevertheless, in more than two decades of having a worm bin in my home, I have found that the worms and associated microorganisms can handle some meat in a worm bin. I do bury chicken bones, for example. If I dig too soon into the pocket of bedding containing the bones and decaying meat, the odor is bad. If I don't disturb it, I don't notice it. When I harvest the castings after several months, what remains is crumbly vermicompost that smells like damp, rich earth and that contains darkened, well-picked bones.

Fellow long-time vermicomposters have had similar experiences. One worm grower I heard from buried the bones from a community chicken barbeque in large outdoor worm bins and said that it took only three weeks for the bones to be picked clean. Dr. Daniel Dindal of the State University of New York at Syracuse suggests adding a good carbon source (such as sawdust or extra bedding) to meat and bones to speed up decomposition time. He finds that if meat is chopped, ground, and thoroughly mixed with the carbon source, rodents won't even be a problem. "I do this successfully all the time in outdoor piles," he says. Several large-scale projects in India use vermicomposting to transform chicken processing waste into valuable natural fertilizer. More recently, studies in India used vermicomposting to break down chicken feathers. Decomposition of feathers usually takes more than five years. Experiments with cow dung and worms composted the feathers in less than three months.

ADD SMALL AMOUNTS AND PROVIDE COVER

The examples above indicate that some meat and bones can be successfully composted if sufficient cover is provided. There are some advantages to putting some of these nitrogen-rich materials into your worm bin. Worms require nitrogen in a form they can use. Nitrogen is also required by the microorganisms that do much of the composting and that are, in turn, eaten by the worms. Since meat contains protein, built from nitrogenous components, eliminating all meat from the system could result in a nutrient deficiency for the teeming organisms that constitute a home vermicomposting system. A further advantage of adding some meat is that more plant nutrients will be in vermicompost produced by worms that have consumed a greater variety of materials. Finally, putting meat scraps into your bin means you don't have to figure out another way to dispose of them.

My personal feeling about burying bones and meat waste in a worm bin is that small amounts are all right. When I clean out my worm bin every six months to a year, I gather the bones into a net bag and hang them in the garage. The next time I clean out the bin and gather more bones, I process the old ones — now completely dry and brittle — by pounding them on concrete with a sledgehammer. These pulverized bones are added to my garden, where plants benefit from their nutrients without my having to purchase bonemeal for nitrogen, potassium, and phosphorus. And centuries from now, the archaeologists excavating my homesite won't have a clue that I was a meat eater!

For the small system inside my home, I use my own judgment on how much meat and bones to bury. I put more of my meat, bones, and dairy wastes into my Patio Bench Worm Bin, away from my immediate living quarters. My advice to you is either avoid placing meat, bones, and dairy products in your worm bin or experiment cautiously with these high-nitrogen materials. Learn for yourself what your system can take within the design and demands you place upon it.

NO-NO'S

Since what is obvious to some of us isn't always obvious to everyone else, let me suggest things that don't belong in a worm bin: plastic bags, bottle caps, rubber bands, sponges, aluminum foil, and glass. Such nonbiodegradable materials will stay there seemingly forever. They will clutter up your developing vermicompost and make it look more like trash. I have seen the same red rubber band over a three-year period in a large outdoor pit!

CAUTION: Don't feed your worms very salty foods. They breathe through their skin and they need to stay moist. Salt will pull moisture from their bodies and can kill them.

Pet Feces

Dog, caged bird, cat, and potbellied pig feces do not belong in the worm bin. The manure of these animals can harbor pathogens and parasites that can be harmful to humans. One example to be guarded against is letting a cat use your worm bin as a litter box. First of all, cat urine would soon make the odor intolerable. Second, the ammonia in the urine could kill your worms. But the greatest concern with cats has to do with a parasitic disease organism, *Toxoplasma gondii*, that can be carried in their feces. Tiny cysts of this protozoan can be inhaled by people and hidden in human tissues. Frequently no outward symptoms occur in the infected person, but it is possible that a pregnant woman could transmit the disease toxoplasmosis to her fetus. Although most cats do not harbor this organism, any cat owner should be very careful in disposing of cat litter. In short, if you have cats, keep them from using your worm bin as a litter box. Although there is evidence that vermicomposting can destroy pathogens, it is best to avoid these feces in a home worm bin.

Some people who raise rabbits build a bin underneath the rabbit hutch for the worms to eat the manure and wasted rabbit feed. If you do this, be sure and add an absorptive bedding to the bin and watch for concentrations of urine. You many need to dilute the urine so that ammonia does not build up.

WASTE-HANDLING PROCEDURES

I use an open stainless steel bowl on my kitchen counter to collect all the organic waste that I will eventually feed to the worms. You can also use any of the commercially available and (decorative) compost holders as long as it allows air to get to its contents and avoids the odors that will soon develop in a tightly closed container (refer to The Secret to an Odor-Free Worm Bin, page 16). I find that adding fresh waste to a tightly closed container becomes objectionable, because it can get pretty "ripe." I am also concerned that if I were to add the contents of the closed container to my worm bin, I would be introducing a large quantity of anaerobic bacteria to the system. Consequently, it would be more difficult to maintain the aerobic conditions we strive for in a worm composting system.

Feeding Frequency

About twice a week, I empty the contents of the holding container into my worm bin. If I have a lot of waste, I empty it more often; if I don't have much, less often. In other words, I don't concern myself with seeing to it that the worms are fed daily, twice a week, or even weekly. My needs, not the worms', dictate how often the worms get fed.

Waste Location

Because I keep records of how much food waste I bury in my worm bins, I weigh the food waste. As I record the weight, I also check my record sheet to determine where I buried food scraps

Burying food waste in different spots (and recording burial spots) helps keep the worms actively eating throughout the bin.

the last time. I rotate around the box, placing food waste in different areas in sequence as illustrated on page 81.

How do you feed your worms? Or how do you place your food waste in the bin? My bin has about nine locations where I can bury the food waste before I have to reuse a spot. Since I bury scraps about twice a week, four and a half weeks pass before I have to dig into a region already containing food waste. By then, much of it is no longer recognizable, having been consumed by the worms or having been broken down by the other natural decomposition processes caused by worm associates in the box.

Burying the Waste

I cover the newly deposited waste with 1" or 2" of bedding, adding more bedding frequently. Covering it makes the food waste less accessible to flies to lay their eggs, and adding bedding adds more carbon sources for the worms. (The alternative method of wrapping food waste in newspaper means it's already covered with bedding.) Then I close the lid or replace the sheet of plastic I have lying loosely on top to retain moisture. With that, I'm through! The whole process takes maybe two minutes, if I take the time to poke around looking for cocoons or baby worms.

The worms will tend to follow the waste, but not necessarily when it is fresh. Food waste will undergo many changes as different kinds of microorganisms invade tissues, breaking them down and creating an environment for other kinds of organisms to feed and reproduce. The worms undoubtedly consume some of the cells from which these tissues are made, but the worms feed also on the bacteria, protozoa, and fungi that thrive in this moist, warm, food-rich environment. Although this book is titled *Worms Eat My Garbage*, I must acknowledge that springtails, sow bugs, bacteria, protozoa, and fungi eat my garbage, too. The worms are there because they help keep conditions aerobic and therefore odor-free, reduce the mass of material to be processed, and produce castings far richer than mere compost. Yet worms don't do the job alone.

Handling Techniques

Techniques vary for handling food waste. One vegetarian who uses worms to process her kitchen waste has large quantities of peelings, wheatgrass roots, and pulp from juicing carrots, celery, and other vegetables. Her worm container is a galvanized garbage can with aeration holes drilled in the lid and top half of the sides. She adds waste daily, merely lifting the lid and dropping the waste onto the surface of the bedding. Occasionally, she throws a double-handful of coir on top of the mass when the lid is lifted. Masses of worms can be seen feeding on the recently deposited waste. Odor has not been a problem in this system. The worms have reproduced greatly, and the end product appears to be well-converted vermicompost.

Should you grind the food waste? No, not for most home systems. Eventually, any soft food waste will break down to become vermicompost, even citrus and melon rinds. I have mentioned that I do pulverize eggshells with a rolling pin to reduce the size of their pieces. There is no question that worms can eat ground food waste more readily than large particles of food. A worm's mouth is tiny, and it has no teeth to break down food particles. However, sometimes the finely ground materials will become anaerobic and putrefy, distressing your worms.

Part of my rationale for using worms inside the home to process food waste is to reduce dependence upon technology. The energy required to grind food waste, dilute it with water, and flush it down the drain, as well as for processing it at the wastewater treatment plant, can be better used elsewhere. For me, to regularly grind food waste before feeding it to worms is inconsistent with why I use worms in the first place.

AVOID OVERFEEDING

"Can I put too much food waste in the worm bin?" This is a common and appropriate question. The answer is yes. You may have a greater than normal quantity of food waste during holidays or harvest activities such as canning or bottling. If you deposit all of it in your worm bin at once, you may overload the system. When this happens, anaerobic conditions will likely develop, causing odor. The first thing I would do to eliminate odor is aerate the bin by turning the material. I would also add fresh bedding. If you can leave the overloaded system long enough without adding any fresh waste, the problem will usually correct itself. This does present you with the problem of how to dispose of your normal quantity of food waste during the interim.

A possible approach to the "overload" times is to set up a separate container with fresh bedding and use a half bucket of vermicompost from your original bin to inoculate this new container with worms and microorganisms. This bin could be maintained minimally, feeding the worms only on the occasions when your week's food waste far exceeds the amount for which your main bin was constructed.

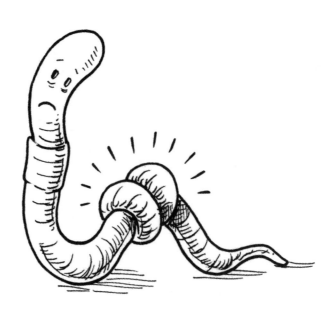

If you have an excess of food waste, use a secondary bin or bucket prepped with a few handfuls of worms and vermicompost.

I have used an old leaky galvanized washtub as a "worm bin annex" that I kept outside near the garage. During canning season, the grape pulp, corncobs, corn husks, bean cuttings, and other fall harvest residues went into this container. It got soggy when it rained, and the worms grew huge from all the food and moisture. We brought it inside at about the time of the first frost. The worms kept working the material until there was no food left. After six to eight months, the only identifiable remains were a few corncobs, squash seeds, tomato skins, and some corn husks. The rest was an excellent batch of worm castings and a very few hardy, undernourished worms.

In other words, given enough time, practically any amount of organic material will eventually break down and decompose in a worm culture. When you want to add fresh material every week, as you do in a system being used to dispose of kitchen waste, there are limits to what is reasonable to add at one time. Your nose is probably the best guide as to when that limit has been exceeded.

Tender loving care for worms basically means respecting them as living organisms. Worm workers provide them with the proper environment and nutrition, check them occasionally, and leave them alone.

The less you disturb them, the better off the worms will be, even though you make some observations of what goes on in their box. Once your worm bin is set up with bedding of the proper moisture content, several sheets of damp newspaper or a sheet of plastic lying loosely on top will retain that moisture. Daily care is unnecessary.

One favorable aspect of having "worms as pets" is that you can go away without having to make boarding arrangements with the vet or a neighbor. You can go away for a weekend, a week, even two weeks, and not worry about your worms. However, if you plan to leave for a month or more, or plan to turn off the heat while away on a winter vacation, you should probably board them while you are gone.

KEEPING TRACK OF THE WORMS

Burial of food waste, whether it is done weekly or more often, consists merely of pushing bedding aside to create a large enough pocket to contain the food waste, depositing the food waste, and covering it with an inch or so of bedding. Train yourself to make a few observations at these times. Does the bedding seem to be drying around the edges? Where are the worms congregating? To find out, you will have to push bedding aside in areas where you have deposited food waste. You can use your hands to do this, or you may prefer to use a trowel or a small hand tool similar to what I call my "worm fork." A worm fork is less likely to injure worms than a trowel.

Sometimes you will see masses of worms feeding around something that especially appeals to them. For curiosity's sake, you might want to note their preference. My worms, for example, love watermelon rind. I place the rind, flesh side down, on the surface of the bedding. Within the next two days, masses of worms of all ages congregate underneath the rind. Within three weeks, all that remains is the very outer part of the rind, looking a lot like a sheet of paper. The same is true for cantaloupe, pumpkin, and squash. Some worm workers have seen worms devouring fresh onion. Odor, if any, won't last long.

There are many other things you can observe. Do older and younger worms prefer different food? When do you first find

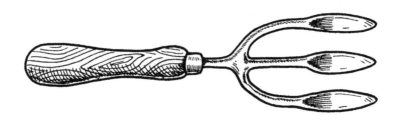

Taking Care of Your Worms

88

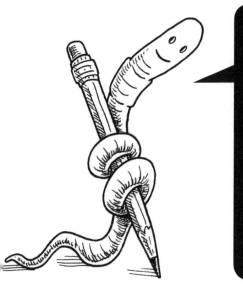

Record-Keeping

I mentioned previously that I keep records of my worm bin activities. In fact, my records that span 25 years provided much of the information in this book. Some of you will want to keep records also, although this would be a distasteful chore for others. If you decide to keep records, it will help to have a utility scale for weighing the food waste, a thermometer for determining bedding temperature, and a pH meter for checking acidity. I currently use a data sheet similar to the one in appendix A.

cocoons? Are they deposited on top of or throughout the bedding? Are any worms mating? Do you see differences in the degree to which the clitellum is swollen?

The preceding questions barely suggest the rich learning experience a home vermicomposting system can provide. Children are fascinated by worms. Many will find the system is an ideal science project. Even a three-year-old was able to understand the concept of feeding food waste to worms. She asked her mother, "Mommy, do I throw this in the garbage can or do I feed it to the worms?"

SOLVING PROBLEMS WITH EXCESS WATER

Plastic worm bins in damp locations, such as basements, tend to accumulate excess moisture in the bottom of the bin. This water comes from several sources. You added water to the bedding when you first set up your bin. Second, you add more water every time you add food waste, because 80 to 90 percent of food waste is water. The third source, however, comes from the water produced by the microorganisms and worms: as they break down waste, they give off carbon dioxide and water as end products. The water vapor produced during these metabolic processes condenses on the smooth, nonporous walls of the plastic container. This condensation water picks up dissolved nutrients as it trickles down through the vermicompost to the bottom. Sometimes referred to as "castings tea" or "vermicompost tea," the actual name for this liquid is leachate. It clearly is not for human consumption. Following are several ways to solve the problem of excess moisture, especially in plastic bins.

Drain and Catch

The bottoms of some plastic bins have drainage holes or an open mesh. A tray underneath captures the water that drains through. This seems reasonable, because no one wants drainage water flowing onto the basement floor. It has a potentially unfavorable consequence, however, as pointed out by environmental toxicologist Dr. Michael Bisesi.

When Bisesi compared the effectiveness of two types of plastic worm bins, he expressed reservations about the system design that relied on a catchment tray to hold the dark water that drained through the vermicompost. His concern was that the nutrient-rich broth could provide ideal culture conditions for many organisms not subject to the controls inside the worm bin. Unchecked, these cultures could proliferate large numbers of bacteria and molds in the open, where "yuck" would be a typical response. Open trays like this would not be considered acceptable in most classrooms.

Add Dry Bedding

I have found that adding dry shredded paper to the surface of the worm bin every two to three weeks helps reduce excess moisture problems. As the vapor pressure inside the bin increases, the water vapor condenses on the lid and "rains" on the new, dry bedding, making it damp. Within a few days, the excess moisture distributes itself throughout the bin, and water standing in the bottom is less of a problem. The worms seem to like the oxygen-rich layer near the fresh bedding on the surface, and I frequently find many of them there. Regular additions of the carbon source that bedding provides also seem to improve functioning of the whole system.

Draw Off with a Turkey Baster

Much of the liquid at the bottom is waste excreted by the worms and not good in their living space. Some people collect this excess liquid so they can apply diluted quantities of it to their houseplants. A turkey baster comes in handy to perform this task. I have found that it's easier to use the turkey baster when I press a strainer into the soggy bedding in the bottom of the bin. Leachate seeps through the strainer and is easily drawn up into the turkey baster without having bedding clog the baster's opening.

Pour Off or Drain

Tilting the bin and holding the bedding back while you pour off the leachate is possible, but this approach may require two people to do the job. If the previous techniques don't work, and your bin is in a location where excess water is a continual problem, I would recommend drilling one hole near the bottom of one side of the bin and plugging it with a cork or rubber stopper. On the few occasions when you have to drain it, just remove the plug. Of course, bins with spigots in the bottom have already made draining leachate easy to do, although you may have to poke something inside the spigot occasionally to maintain the opening if it gets clogged.

WORM AND WORM BED MAINTENANCE

In about six weeks, you may begin to see noticeable changes in the bedding. It will get darker, and you will be able to identify individual castings. Although you add food waste regularly, the bedding volume will slowly decrease. As more of the bedding and food waste is converted to earthworm castings, extensive decomposition and composting by other organisms in the bin also take place. I mentioned earlier that the proportion of castings increases as the environmental quality for your worms decreases. There will come a time when so much of the bedding in the box becomes castings that the worm population will suffer. Because each system is different — depending on bedding used, quantity of worms, types of waste fed to them, bin temperature, and moisture conditions — it is not possible to predict precisely when you must deal with changing the environment of your worms. It is important, however, to get them away from their castings and to prepare fresh bedding for them at the right time.

Your particular goals, described in chapter 2 in terms of whether they require high, low, or medium levels of maintenance, will help decide this. For example, to harvest extra worms for fishing, you will have to change bedding more frequently. Plan on doing so every two to three months, and figure that it is a high-maintenance system.

Harvesting Worms

If you don't want to harvest worms from partially decomposed food waste and bedding, but do want high-quality vermicompost almost fully converted to worm castings, the trade-off is losing your worm population. Accepting this trade-off means adapting your own behaviors to complement this loss. In northern systems, for example, you might bury food waste in your bin for the four winter months, and then let it sit unattended for another three to four months. By July you will find a bin full of fine, black worm castings, but there will be very few worms remaining — perhaps not more than a dozen. These fine castings can be used as topdressing on your houseplants and in your garden for a late shot of nutrients. This was referred to earlier as "the lazy person's technique" for maintaining a worm bin. I've done it, and it does work. When I'm using this system, though, I have to compost food waste in outdoor compost piles during spring, summer, and fall.

High- and medium-maintenance systems require that you harvest the worms, or at least give them the opportunity to move into fresh bedding. For a high-maintenance system, plan to do this every two to three months, while medium-maintenance means going about four months before you take action. Your first-time harvest of a 2 × 2-foot bin takes two to three hours, but it goes faster when you gain experience. If you have curious friends or family to help, even faster harvests are possible.

The following pages show several methods you can choose from to harvest your worms and the vermicompost.

Harvest Method #1

DUMP AND HAND SORT

Vermicompost from this sorting process will vary in consistency, depending upon how long the bin has been going, how much and what kind of waste was buried, and how much decomposition has occurred. Some of the most recently buried food waste can be put right back into the fresh bedding. See page 96 for an illustration of the process.

NEEDED

- Large sheet of heavy plastic or tarp

- Lamp or bright source of light

- Fresh bedding (see chapter 4)

- Plastic dishpan or other container for worms

- Plastic or metal garbage can or heavy-duty plastic bag for vermicompost

1. SPREAD A LARGE PLASTIC SHEET or tarp on the ground, floor, or table, and dump the entire contents of the worm bin on it.

2. MAKE ABOUT NINE PYRAMID-SHAPED PILES IN A WELL-LIT SPACE. You should see worms all over the place. If the light is bright enough, they quickly move away from it and head toward the center of each vermicompost pile.

3. LEAVE THE PILES ALONE FOR FIVE TO TEN MINUTES TO GIVE THE WORMS TIME TO RETREAT TO THE NEW PILES YOU HAVE MADE. While you're waiting, you can prepare fresh bedding and restock the empty worm bin.

4. RETURN TO THE PILES AND GENTLY REMOVE THE OUTER SURFACE OF EACH ONE. As you do so, worms on the newly exposed surface will again react to the light and retreat toward the interior. By following this procedure one pile at a time, you will find that when you return to the first pile, the worms will have disappeared again, and you can repeat the procedure. The process may take about an hour for an 18- to 20-gallon bin.

Eventually, the worms will aggregate in a mass at the bottom of each pile. Remove the vermicompost that collects on top of them, and put the worms in the container you have ready for them. Only if you are going to weigh the worms do you need to remove all vermicompost from this batch.

5. WHEN THE WORMS ARE SORTED AND WEIGHED, ADD THEM TO THE TOP OF THE FRESHLY RE-BEDDED BIN. You are now ready for another cycle.

WET COMPOST

Vermicompost from plastic bins may be excessively moist. If water drains onto the plastic sheet, or if the pile slumps down from excess water, you know it is too wet. Sorting through this sticky material is not pleasant, and it is very difficult to get the worms out.

I have placed such soggy vermicompost in a heavy-duty corrugated container and let it sit in a dry place for several weeks, until the vermicompost was well stabilized and earthy-smelling. The worms I found were tiny, however, and could obviously have benefited from some new food and drink.

SAVING THE WORM COCOONS

A large number of cocoons and baby worms should be present in the vermicompost from which worms were harvested. If you wish, you can save many of them by letting the vermicompost sit for about three weeks. Then attract them with a long, narrow strip of food. This may be one occasion where use of a blender is appropriate: Make a slurry of food waste, perhaps with some oatmeal, cornmeal, or other grain mash in it. With your fingers or a trowel, make a groove down the center of the vermicompost and fill this groove with the slurry. In a couple of days, you should be able to remove concentrated batches of young worms from underneath this narrow strip. Repeat two or three times to obtain new hatchlings you can add to your regular bin.

DUMP AND HAND SORT, AT A GLANCE

Dumping and hand sorting is an effective way to separate worms from their castings.

1. WORMS AND VERMICOMPOST

PLASTIC SHEET OR TARP

6' APPROX

6'

MAKE CONE-SHAPED PILES.

2.

BRIGHT LIGHT

EACH PILE CONTAINS WORMS AND VERMICOMPOST.

3. WORMS GO TO THE BOTTOM OF EACH PILE TO AVOID LIGHT. REMOVE TOP AND SIDES.

4.  AFTER REMOVING
VERMICOMPOST YOU
WILL FIND MASSES
OF WORMS AT
THE BOTTOM OF
EACH PILE.

PLACE WORMS IN
A CONTAINER AND
WEIGH THEM.  **5.**

6. SAVE VERMICOMPOST
FOR GARDEN AND
HOUSE PLANTS.

VERMICOMPOST

7.  ADD WORMS
TO BOX WITH
NEW BEDDING.

Harvest Method #2

LET THE WORMS DO THE SORTING

Letting the worms do their own sorting is a slower but simple process. When the bedding has diminished to the extent that it is not deep enough to make a hole to bury fresh food waste, it is time to add fresh bedding.

PREPARE ABOUT HALF THE ORIGINAL QUANTITY OF FRESH BEDDING. Pull all of the vermicompost in your bin over to one side, and add the new bedding to the empty side. Bury your food waste in the new bedding, and let the worms find their way to it. It is helpful to replace the plastic sheet only on the side with the fresh bedding to permit the other side to dry out more rapidly.

EVERY TWO TO THREE MONTHS, YOU CAN REMOVE THE VERMICOMPOST, replace it with more fresh bedding, and keep going back and forth from one side to the other in this manner. The vermicompost you remove will still have some worms in it, but enough should have migrated to the new bedding that you needn't worry about harvesting the few that remain.

Some commercial bins use an effective variation of this self-sort technique by placing a vertical mesh or perforated screen between the two halves of the worm bin. Worm workers place food waste in one side for a period of time, then use the other side. The worms freely move horizontally through the mesh or screen holes on their own time. Eventually, the side with the fresher material has more worms, and the older side has stabilized vermicompost, which can be removed.

Continuous flow and tiered systems improve upon the horizontal method of letting the worms do the sorting by having the worms move up toward the new supply of food, leaving their castings below. Our experience with this vertical system is that the lower trays or lower part of the bag become full of finely processed castings with no recognizable food waste. The lower trays may also have worms. It's easy to sort the worms from the well-processed castings by picking them out or dumping the tray. We have found that not all of the worms make it to the upper levels where the fresh food is.

1. PULL VERMICOMPOST AND WORMS TO ONE SIDE OF THE BOX.

2. ADD NEW BEDDING TO VACANT SIDE.

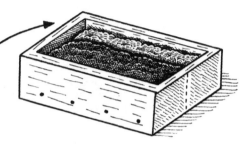

3. BURY GARBAGE IN NEW BEDDING. WORMS MOVE TO NEW BEDDING IN SEARCH OF FOOD.

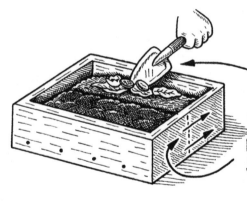

4. REMOVE VERMICOMPOST IN 2-3 MONTHS AND REPLACE WITH NEW BEDDING.

BLACK PLASTIC ON FRESH SIDE ONLY

Harvest Method #3

DIVIDE AND DUMP

Still another method for harvesting worms is the divide-and-dump technique. You simply remove about two-thirds of your vermicompost and dump it directly onto your garden's surface. No digging or turning; no muss, no fuss. Add fresh bedding to the vermicompost still left in the box. Enough worms and cocoons usually remain there to populate the system for another cycle. If you're concerned about invasive worms, you should freeze the vermicompost to eliminate any worms before dumping it.

1. TAKE OUT ALL BUT ⅓ OF WORMS AND VERMICOMPOST.

ADD NEW BEDDING.

SAVE ⅓

2.

ADD VERMICOMPOST TAKEN FROM BOX TO THE GARDEN... WORMS AND ALL.

Additional Containers

Many people find it most convenient to have more than one worm bin, one active and the other resting. When they can no longer find a suitable place in the first worm bin to bury food waste, they let it sit. They merely set up another bin with fresh bedding and soil, move some worms over from the first bin, and use the new bin exclusively. Meanwhile, the worms and microorganisms remaining in the first bin continue to process the waste, which eventually stabilizes to consist of well-processed castings.

SCREENING METHODS

A number of different worm-screening devices can be constructed using 1/8" or 1/4" hardware cloth. The simplest method is to make a frame around the cloth, place it over a container, and then sift the vermicompost into the container. Anytime you sift the vermicompost, it needs to be crumbly and fairly dry. Some worm workers place the cloth over a cardboard cylinder or cut out the side of a 5-gallon bucket and put the cloth over the opening. Whatever method you use, it is best to start with a larger screen and then move to a smaller one. I have even seen one worker hold a large colander over a bucket.

Another method is to put an old screen on top of the vermicompost and start feeding and placing bedding on it. The younger small worms will go up first. Larger worms will stay below and continue to feed until they either run out of food or find that the vermicast is uncomfortable to them. They will then squeeze through the screen to eat. Once they are on top of the screen, you can lift it off and harvest the castings below.

Harvesting can also be done with a mesh bag that onions and oranges are sold in. Let your worms get hungry by not feeding them for ten days. Then put their favorite foods (like a banana or ripe apple) in the bag and bury it in the bin. Within a day or two the worms will move through the holes in the bag and start eating the food. You won't get all of the worms in the bag, but you will get a lot.

The maintenance system you choose will depend upon your preference, your lifestyle, and perhaps your schedule at the time. You may find yourself using all of these systems at various times. At any rate, maintaining your home vermicomposting system can be a flexible process and is really very simple.

TEMPERATURE EXTREMES

We've already discussed the need to maintain a worm bin temperature that will permit the worms to thrive. What if you live where it gets too cold for the worms in the winter or too hot for them in the summer? Three approaches for each extreme are feasible. Let's consider cold winters first.

Winter Methods

BRING THE BIN INDOORS. If it's warm enough inside your home for you, it's warm enough for the worms. Older people who prefer an indoor winter temperature of 80°F (27°C) or higher may find their worms prospering. Most people prefer to keep their worm bin in the basement. Apartment dwellers and those with no basements make room for their worm bin in their kitchen, utility room, or even a living room! In a basement colder than 40°F (4°C), most worms will live but move and eat very slowly. The worms and their associated composting organisms will not process as much food waste as if they were closer to their optimum temperature. Use of a 7-watt night light within the bin should bring the temperature up, although the light will inhibit their activity. The worms can work in the dark under a few sheets of newspaper.

INSULATE YOUR OUTDOOR BIN. To meet the demands of cold Canadian winters near Toronto, Sam Hambly, a worm grower, enthusiast, and educator, designed a 24" × 48" × 48" worm bin. The sides were plywood panels held together with screen door hooks and eyes, and the open bottom gave worms access to the soil beneath. (Or the worms beneath had access to the goodies above!) Styrofoam provided insulation in the form of four sheets of 2" pieces cut to fit inside the plywood walls.

Sam built a plywood lid with strips of wood along the edges that helped hold the unit together when in position. One of the most important design elements was a thick, supplemental foam lid. Sam cut the lid about 4" smaller than the inside opening for placement directly on a plastic vapor barrier sitting on the bedding. The space around the sides allowed air to get to both the decomposing materials and the worms. He put leaves and garden residue in addition to food waste in this large bin. The large volume of organic materials inside the bin generated heat from its composting, and the floating lid retained enough of this heat to keep the worms warm and active even during the coldest parts of winter.

ADD SUPPLEMENTAL HEAT. For years I have set up worm bins either in my garage or outdoors, trying to find the right combination of insulation and supplemental heat to keep my worms from freezing during our southern Michigan winters — which can reach 15°F below zero (–26°C). I now place bales of straw around my Patio Bench Worm Bin and lay a 1" Styrofoam sheet inside the lid for insulation. I then insert a birdbath water heater with an electric immersion coil into a 2-gallon (7.6 L) jug of water and place the jug in the center of the worm bin. I wrap the connection between the heating coil and extension cord carefully with a plastic bag and electrical tape to prevent moisture from getting into the connection. This unit is plugged in throughout the winter. The thermostat turns the unit on when the temperature goes below 40°F (4°C) and maintains the water at that temperature. Another option is a fish-tank heater.

Even though the outside edges of the worm bin may freeze, I can always find live worms near the central core containing the water. A thermostat set at 65°F (18°C) would produce more active worms but would also consume more electricity. My main waste disposal in winter is my basement worm bin, so the outside "annex" doesn't receive much food waste. In our household of two, we definitely do not produce enough food waste to keep a worm bin the size of Sam Hambly's going during winter.

Summer Methods

BRING THE BIN INDOORS. Temperatures above 86°F (30°C) will start to cause problems. At 95°F (35°C), the worms are vulnerable to overheating and could die. At that temperature the worms have a much higher demand for oxygen, so they will deplete their bedding of oxygen much more quickly than at cooler temperatures. If some of the worms die, their bodies quickly begin to decompose, which removes even more oxygen. Some people find the easiest solution is to bring the worm bin indoors, where it can be kept out of the sun.

KEEP THE BIN IN THE SHADE. Often it is enough to place the worm bin under a tree where there is enough shade to keep direct sunlight from bearing down on the bin. Placement in a garage or under a shed roof also may work. If the exterior of your worm bin is a dark color, you may lessen absorption of heat by taping aluminum foil to the top of the bin to reflect heat away from the bin itself.

USE EVAPORATIVE COOLING. Some people drape a burlap bag around their worm bins, wet it, and lower the temperature inside by moving air across the wet cloth. A breeze is best, but use a fan if necessary. The process of evaporating water from the cloth removes heat from the bin in the same way that we cool down when we wear a wet shirt. Often, the period of critically high temperatures is relatively brief, so one doesn't have to take extraordinary measures like this for very long to keep worms healthy.

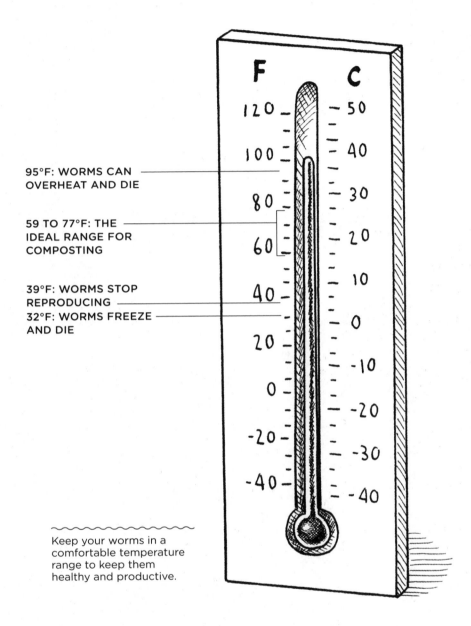

F

120 — — 50

100 — — 40

95°F: WORMS CAN OVERHEAT AND DIE

80 — — 30

59 TO 77°F: THE IDEAL RANGE FOR COMPOSTING

60 — — 20

— 10

39°F: WORMS STOP REPRODUCING

40 —

32°F: WORMS FREEZE AND DIE

— 0

20 —

— -10

0 — — -20

-20 — — -30

-40 — — -40

C

Keep your worms in a comfortable temperature range to keep them healthy and productive.

Are you one of the thousands of people who have mixed reactions toward worms? Do you feel revulsion toward these moist, wriggly creatures at the same time that you are fascinated by them? Are you somewhat curious but don't want to learn too much about them? Then this chapter is for you. If you'd like to learn even more, refer to the resources on page 176.

Can a worm see?

Contrary to the popular cartoon image of worms, they have no eyes and cannot see. They are, however, sensitive to light, particularly at their front ends. If a worm has been in the dark and is then exposed to bright light, it will quickly try to move away from the light. A nightcrawler, for example, will immediately retract into its burrow if you shine a flashlight on it some wet spring night.

The sensory cells in a worm's skin are less sensitive to red light than to light of mixed wavelengths. If you want to observe worms under less intrusive conditions, you can take advantage of this fact by using a red-light flashlight or headlamp, commonly used for night vision. Your eyes will adapt to the lower level of light these aids provide, and the worm will move more naturally than it does under bright light.

Where is the worm's mouth?

A worm's front and back ends are more technically known as anterior and posterior. The mouth is in the first anterior segment. A small, sensitive pad of flesh called the prostomium protrudes above the mouth. When the anterior end of the worm contracts, the prostomium is likely to plug the entrance to the mouth. When the worm is foraging for food, the prostomium stretches out, sensing suitable particles for the worm to ingest. I was amazed at how wide a nightcrawler can open its mouth when I first saw it on video. The worm curled its anterior segments upward, revealing a wide-open mouth for a fraction of a second. Later, I saw the worm grab a leaf with its mouth and drag it toward its burrow. In *The Formation of Vegetable Mould through the Action of Worms, with Observations on Their Habits*, Charles Darwin documented that not only do these worms pull leaves into their burrows, but 80 percent of the time they draw them in by their tips. He experimented with paper triangles and artificial sets of pine needles. The majority of the time, the worms pulled the material into burrows in a "thoughtful" manner. Although it has not been proven, it appears that worms may be fairly intelligent beings.

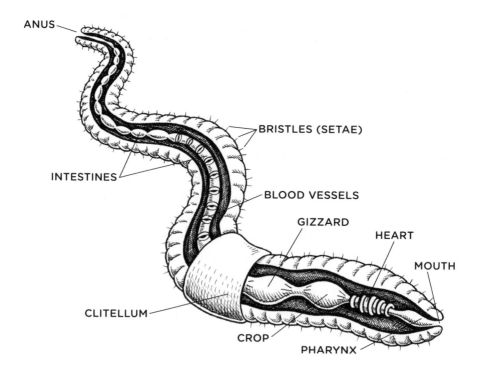

ANUS

BRISTLES (SETAE)

INTESTINES

BLOOD VESSELS

GIZZARD

HEART

MOUTH

CLITELLUM

CROP

PHARYNX

Does a worm have teeth? How do they eat?

No teeth. The mouth and pharynx are highly muscular, but they do not contain teeth. Food enters into the mouth, where the pharynx pushes it into the esophagus. Glands then release calcium carbonate to rid the worm's body of excess calcium. From here the food is stored in the crop until it goes into the gizzard. The worm does not have a stomach.

How does a worm grind its food?

Because worms have no teeth, they have little capacity to grind their food. They are limited to food that is small enough to be drawn into the mouth. Usually this food is softened by moisture or by bacterial action. Bacteria, protozoa, and fungi help break down the ingested organic material. Every worm has one muscular gizzard, however, which functions similarly to gizzards in birds. Small grains of sand and mineral particles lodge in the gizzard. Muscular contractions in the gizzard wall compress these hard materials against each other and the food, then mix it with some fluid, and grind all into smaller particles. One reason for manually spreading a handful of topsoil, rock dust, or lime into worm bedding is to provide worms with small, hard particles for their gizzards.

What happens to food once it leaves the gizzard?

The ground-up food enters the worm's intestine, where modifications occur to the decaying material and the microorganisms to chemically break down molecules of food nutrients. These simplified nutrients then pass through the intestinal wall for absorption into the bloodstream, and are carried where needed. Undigested material, including soil, bacteria, and plant residues, passes out of the worm through its anus as a worm casting.

Do worms have hearts?

They have five pairs of aortic arches that function like a heart. They pump blood to the front of their body through the dorsal blood vessels and to the back of their body through the ventral blood vessels.

What about a brain?

Worms have a group of nerve cells called a ganglion, which acts as a brain. Connected to a nerve cord that runs the length of the body, ganglia branch off in each segment. Although it is a simple system, it is sensitive to a number of things — including light, temperature, and vibrations.

Do worms need air?

Worms require gaseous oxygen from the air. The oxygen diffuses across their moist skin tissue from the region of greater concentration of oxygen (the air) to that of lower concentration (inside the worm). When water has been sufficiently aerated, worms have been known to live under water for a considerable length of time.

Carbon dioxide produced by the bodily processes of the worm also diffuses through its moist skin. Moving from higher concentration to lesser concentration, carbon dioxide transfers from inside the worm's body out into the surrounding bedding. A constant supply of fresh air throughout the bedding helps this desirable exchange of gases take place.

How do they move?

On each segment or section there are bristles called setae. As setae anchor the worm to the ground, other sections will contract or expand, causing the worm to move.

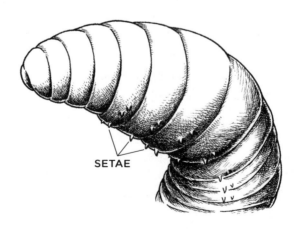

SETAE

If a worm is cut in half, will both parts grow back?

This depends on the type of worm, where it was cut, and how much it was damaged. Studies have shown that *Eisenia fetida* can regenerate a head, but if many segments are lost, the chances decline. So it is rare that a tail will produce a head. Almost all segmented worms are able to grow a new tail.

On rare occasions you may find a worm with two tails, both at the same end. This condition can be caused by injury to the worm in the posterior end, which results in growth of a new tail adjacent to the original tail.

Do worms die in the bin?

Worms undoubtedly die in any home worm bin, but if your box is properly maintained, you rarely will see a dead worm. Their bodies quickly decompose and are cleaned up by the other organisms in the box, leaving few dead worms you can recognize.

If large quantities of worms seem to be dying, you should attempt to determine the cause and correct the problem. Is it too hot? Are toxic gases building up in the bedding that cause the worms to surface and get away? Did you stress the worms by adding too much salty, aromatic, or acid-producing food?

You'll need to make some educated guesses about what the problem is and try to correct it. Adding fresh bedding to a portion of the box sometimes is enough to correct the situation, providing a safe environment toward which the worms can crawl.

How long does a worm live?

Most worms probably live and die within the same year. Especially in the field, most species are exposed to hazards such as dryness, weather that is too cold or too hot, lack of food, or predators. In culture, individuals of *Eisenia fetida* have been kept as long as four and a half years, and some *Lumbricus terrestris* have lived even longer.

Once your home vermicomposting system has been established for a while, you will begin to find creatures other than earthworms present.

This is a normal situation, but it could be alarming if you were brought up to think that all bugs are bad bugs. Most of them are, in fact, good "bugs," and few of them actually are classified biologically as bugs. They play important roles in breaking down organic materials to simpler forms that can then be reassembled into other kinds of living tissue. This whole array of decomposer organisms gives meaning to the term "recyclers." You could spend a lifetime studying the various creatures in a worm bin trying to determine who eats whom and under what conditions.

FOOD WEB OF THE COMPOST PILE

ENERGY FLOWS IN THE DIRECTION OF THE ARROW.

1° = FIRST-LEVEL CONSUMERS
2° = SECOND-LEVEL CONSUMERS
3° = THIRD-LEVEL CONSUMERS

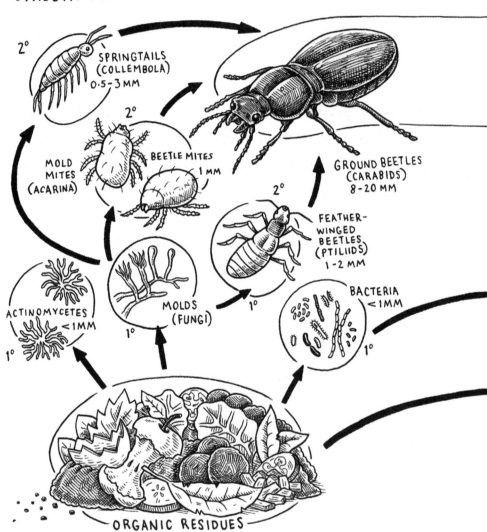

2° — SPRINGTAILS (COLLEMBOLA) 0.5–3 MM

2° — MOLD MITES (ACARINA) — BEETLE MITES 1 MM

GROUND BEETLES (CARABIDS) 8–20 MM

2° — FEATHER-WINGED BEETLES (PTILIIDS) 1–2 MM

ACTINOMYCETES <1MM 1°

MOLDS (FUNGI) 1°

1° — BACTERIA <1MM 1°

ORGANIC RESIDUES

Adapted from Dindal 1971, *Ecology of Compost.*

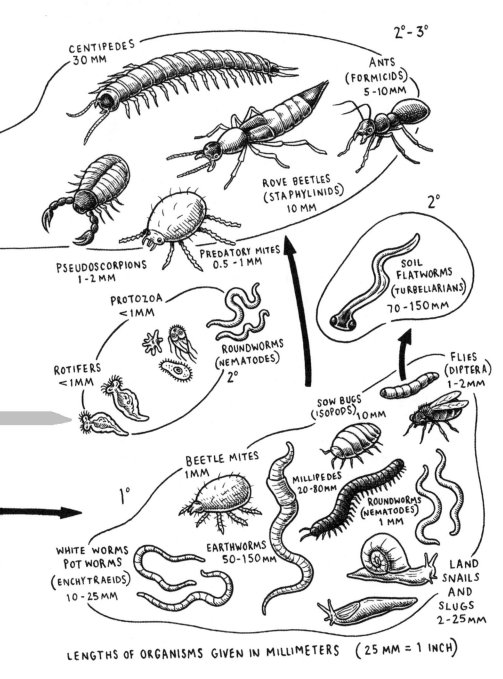

CENTIPEDES
30 MM

2° - 3°

ANTS
(FORMICIDS)
5 - 10 MM

ROVE BEETLES
(STAPHYLINIDS)
10 MM

PSEUDOSCORPIONS
1 - 2 MM

PREDATORY MITES
0.5 - 1 MM

2°

SOIL
FLATWORMS
(TURBELLARIANS)
70 - 150 MM

PROTOZOA
<1MM

ROUNDWORMS
(NEMATODES)
2°

ROTIFERS
<1MM

FLIES
(DIPTERA)
1 - 2 MM

SOW BUGS
(ISOPODS) 10 MM

BEETLE MITES
1 MM

MILLIPEDES
20 - 80 MM

ROUNDWORMS
(NEMATODES)
1 MM

1°

WHITE WORMS
POT WORMS
(ENCHYTRAEIDS)
10 - 25 MM

EARTHWORMS
50 - 150 MM

LAND
SNAILS
AND
SLUGS
2 - 25 MM

LENGTHS OF ORGANISMS GIVEN IN MILLIMETERS   (25 MM = 1 INCH)

A NETWORK OF ORGANISMS

In worm bins, energy flows from organism to organism as one is eaten by the other in a natural recycling system. Millipedes, centipedes, and ants are less likely to find their way to worm bins set up with shredded-paper beddings. Ants prefer a dry environment. If you find them in your bin, the bedding is probably too dry. Millipedes are more likely to be found in a bin where the bedding is composted plant material or manure. In their natural environment, they are scavengers that eat decaying wood particles, leaves, or plants. Centipedes are usually found in moist habitats such as under rocks, in leaf litter, or rotting logs. They are territorial, and should you find one in your bin, there won't be many of them, so they will be easier to pick out and kill.

FIRST-LEVEL CONSUMERS. Organisms that consume waste directly are first-level (1°) consumers. They include microscopic actinomycetes, molds, and bacteria. Actinomycetes are fungus-like bacteria that produce thin filaments radiating from a central point. Their presence gives compost and soil its earthy odor. Earthworms, beetle mites, sow bugs, enchytraeids, and flies are also first-level consumers when they consume waste directly.

SECOND-LEVEL CONSUMERS. These eat 1° consumers or their waste products. Examples of second-level (2°) consumers in a compost pile include springtails, mold mites, and feather-winged beetles that eat molds, bacteria, and actinomycetes. When protozoa and rotifers eat bacteria, they function as 2° consumers. The function an organism serves, however, changes depending on its food source at a particular time. An organism may be a 1° consumer at one time, such as when an earthworm eats a leaf, or it may be a 2° consumer when it consumes the bacteria that cause a piece of apple to decay.

THIRD-LEVEL CONSUMERS. These are flesh eaters, or predators, which eat 1° and 2° consumers. Third-level (3°) consumers in a compost pile or in your worm bin might include centipedes, rove beetles, ants, and predatory mites.

MICROSCOPIC ORGANISMS. You won't be able to see many of the organisms pictured in the illustration on pages 114–115 because they are microscopic (bacteria, protozoa, nematodes, and rotifers). Others, such as the springtails and mites, are so small that you will probably need a hand lens to get a better look. Brief descriptions of the more common "critters" follow.

Pot Worms
(Family Enchytraeidae)
Known commonly as white worms or pot worms, enchytraeids are small (¼" to 1" [0.6 cm to 2.5 cm] long), white, segmented worms. You might mistake them for newly hatched redworms because of their size. However, newly hatched redworms are reddish because of their red blood. Although related to the larger earthworms, enchytraeids do not have a hemoglobin-based blood and remain white throughout their lifetime.

Enchytraeids eat decomposing plant material rich in micro-organisms, but they digest only part of it, just like the earth-worms. This partial breakdown of litter helps make food material available for other decomposers. Their manure provides further sites for other microbial activity.

Pot worms are harmless and tend to thrive in low pH and high-moisture conditions.

Some worm growers incorrectly refer to enchytraeids as nematodes and feel that they should try to get rid of them.

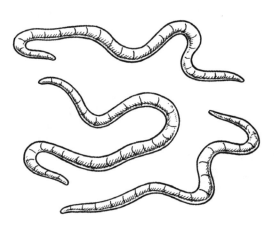

Nematodes, also important decomposers, are undoubtedly present in large quantities in worm bins, but you would not be likely to see them without a microscope. Under the microscope, nematodes clearly are not made up of segments; they are smooth and round all along their length. Some commercial worm growers are concerned that enchytraeids will compete with redworms for feed, and may attempt to control their numbers. Since the purpose of having a home vermicomposting system is to get rid of food waste, the presence of an organism that helps to do the job is an asset, not a detriment. My position concerning enchytraeids is "Let them be."

Springtails
(Order Collembola)
In your worm bin, you may see a sprinkling of hundreds of tiny (1/16"; 0.2 cm) white creatures against the dark background of the decomposing bedding. When you put your finger near them, some spring away in all directions. Springtails are primitive wingless insects with a pointed prong extending forward underneath their abdomen from the rear. They have three body segments, six legs, and two short, stubby antennae. By quickly extending this "spring," known as a furcula, they jump all over the place. Snow fleas are a kind of springtail. Other members in the same order scientists call Collembola do not have the springing tail. Collembola feed on molds and decaying matter. They are important producers of humus and are considered to be among the most important soil organisms.

Springtails are not only numerous, they are diverse, with more than 6,000 species described. They live in all layers and types of soils from Antarctica to the Arctic, and thrive in moist environments.

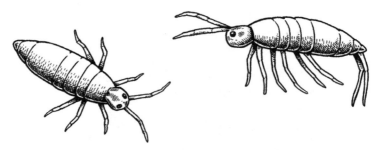

Isopods
(Order Isopoda)

Isopods are easy to identify because the series of flattened plates on their bodies makes them look like tiny armadillos. If one of these gray or brown, ½"-long creatures rolls up into a ball, it is commonly called a pill bug. Its scientific name is *Armadillidium vulgare*. Sow bugs are related to pill bugs but don't form little balls. Other common names for isopods are woodlice, roly-polies, and, in Australia, slaters.

Isopods are crustaceans, related to crayfish and lobsters. They have gills and need a moist environment for the exchange of gases, but they have adapted to life completely on land. The

dampness of a worm bin is ideal for them. If you use manure bedding in your worm bin, you are almost certain to have a few isopods grazing over the surface. They shred and consume the toughest material and are highly beneficial in a worm bin. They won't harm living worms since they eat vegetation and leaf litter, but decaying animal matter is also on their menu.

Centipedes
(Class Chilopoda)

Centipedes are just about the only critters that I kill on sight in a worm bin. You'll probably never have many of them, but they are predators that occasionally kill worms. Centipedes move quickly on their many legs and have a pair of poison fangs on the first trunk segment. Generally, if you find them, they will be on the surface of the bin. Should you capture a centipede and want to

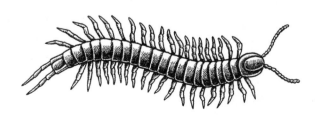

look at it without killing it, immerse the animal briefly in some soda water to let the carbon dioxide anesthetize it.

You can tell a centipede (hundred-legged) from a millipede (thousand-legged) in two ways by looking carefully at their bodies. Centipedes are flattened and have only one pair of legs per segment. Millipedes are cylindrical and have two pairs of legs on each abdominal segment. Both of these segmented creatures are arthropods, not earthworms.

Millipedes
(Class Diplopoda)

You may spot a few millipedes in your worm bin, especially if you use manure, leaf mold, or compost as part of your bedding. They are slow moving and can be found in all layers of the bin. They are vegetarians and won't kill your worms. In fact, they are very helpful and contribute more to breaking down organic matter than generally realized. After leaves have been softened by water and bacteria, millipedes eat holes in them, helping springtails, mites, and other litter dwellers skeletonize them so that only leaf ribs remain. I wouldn't ever consider killing a millipede.

Predatory Planarians
(Class Turbellaria)

Several species of terrestrial planarians prey on earthworms. Planarians are flatworms, or turbellarians. They are normally free-living animals that live in moist places (under rocks, logs, boards, or bricks) rich in organic material. Planarians show no segmentation. Like earthworms, they exchange gases through their moist, mucus-covered skin, which may be bright orange or dull yellow with one or more long black stripes. Several inches long, planarians glide along surfaces with thousands of cilia moving on their ventral surfaces.

A predatory planarian attacks an earthworm by mounting it and attaching a flexible protruding pharynx (feeding tube) to the worm. Digestive juices and secretions from glands in its skin liquefy the worm tissue. The planarian then sucks up the juices through its feeding tube, thereby consuming the earthworm.

Land planarians in the genus *Bipalium*, originally from Southeast Asia, are invasive to the United States. *B. kewense* has been found in greenhouses in 14 states and in natural habitats in at least 7 states. These worms are aggressive hunters and will attack an earthworm 100 times its mass. In some cases, *B. kewense* and *Dolichoplana striata* have been a problem in earthworm-rearing beds in the southern United States.

It is believed that the native New Zealand planarian *Arthurdendyus triangulatus* was introduced to areas outside of that country by way of plant containers moving from nurseries and garden centers. It is now established in Ireland, Great Britain, and the Faroe Islands.

A. triangulatus was first observed in Belfast, Northern Ireland, in 1963. This planarian appears to eat any earthworms; however, most susceptible are earthworms that feed on the surface and have semipermanent burrows. A study in Ireland showed that the biomass of earthworms declined by 20 percent when the area was infested with this planarian, and *Lumbricus terrestris* numbers were reduced by 75 percent.

Although these predators are unlikely to appear in your worm bin, I mention predatory planarians here so that more people will be on the alert for them. If you do find one, do not try to kill it by chopping up. They reproduce by fragmentation, and all you will do is double the number. A recommended solution for killing a planarian is to cover it in salt, vinegar, or citrus oil.

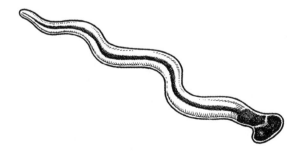

Mites
(Subclass Acarina)

You will undoubtedly have many, many mites in your worm bin. Like the springtails, mites are so small it is difficult to see them except as minute dots moving across the surface of the bedding. Mites have eight legs and a round body. Great diversity exists among the tens of thousands of mite species. Some eat plant materials, such as mold, algae, decaying wood, and soft tissues of leaves. Others consume the excrement of other organisms. Beetle mites don't travel very far on their own, but they travel as stowaways on dung beetles that transport them from one dung heap to another.

The white or brown mite can achieve such high numbers that the worms may refuse to feed. The red mite is a parasite and will harm earthworms and their cocoons. Mites are more likely to be present in very wet beds and may concentrate on one or another kind of food, completely covering the surface. If this happens, remove and burn the mite-infested food, or put it out in the sun to kill the mites. Bait others in the same way by placing a piece of bread or a melon wedge flesh side down on the bedding, then remove it when the mites concentrate on its underside.

To create conditions that aren't so favorable for earthworm mites, leave the cover off your bed for a few days to reduce bedding moisture. Also, do not let your bin become too acidic. This can happen when you overfeed and food begins to ferment and lowers the pH of the system.

Flies
(Order Diptera)

If someone were to ask me, "What is the most annoying problem you have encountered in having a worm bin in your home?" I would have to answer, "Flies. Not odor, not maintenance, not worm crawls, but flies — whether they be fruit flies or fungus

Other Critters and Pests

gnats." Not every box has them, and not every box that experiences them has them all the time, but when they are present, they are a nuisance.

The flies I particularly alert you to are tiny: *Drosophila melanogaster*, commonly known as fruit or vinegar flies. You may remember this organism from genetics lab. There are many families of fungus gnats, all of which are in the order Diptera. These may live in the worm bin, often seeming to sort of jump and scurry around on the bedding. Gnats are not dangerous; however, they bite. Fruit flies and gnats undoubtedly come in on fruit peels and rinds, or are attracted by them in late summer or early fall. When food waste is not buried, bin conditions are favorable for their reproduction. "Prolific" is an apt descriptor. When they land in your orange juice, beer, or champagne, however, they're a bit much. Even the most tolerant guests figure that maybe you've gone too far with this "ecology thing."

There are ways, however, to prevent flies and combat the ones that do show up.

PREVENTION

Always bury food waste. If you don't, the odor will bring flies from far and near. They need exposed food to lay eggs. They will do as they were meant to do — lay eggs in a food source for their young. The fly eggs will hatch into larvae, and the larvae will consume the food source as they go through several larval stages prior to forming a pupa from which will hatch an adult. Fly larvae, or maggots, are quite efficient in breaking down the food waste, but most of us are repulsed by them and would prefer not to see them in our worm bins.

Keep flies away from your storage container. I prefer a screened lid to one that would keep air out of the container because I want oxygen-loving organisms to start the decomposition process from the very start. If flies are loose in your home, you might place your container in the refrigerator until it's time to bury the food waste in your worm bin. Some people even freeze the waste.

When you deposit food waste in your worm bin, take the time to cover it with a layer (1" to 2") of bedding. You may need to add fresh bedding to do this; the additional carbon in the system will be good. Two reasons explain why burying the food waste is necessary. First, the bedding will help to prevent the smell of the food source from reaching the sensitive sensors that all flies have. They will be less likely to know the food source is there. Second, flies are not burrowers. They have no way of burrowing down into the bedding to get to the food sources for laying their eggs. Flies aren't likely to lay eggs on clean newspaper bedding.

CONTROLS

If flies become a problem, try one or more of these remedies:

- Get rid of the adults by trapping near either the worm bin or the food-storage area.

- Stop feeding the worms for two or three weeks to let the existing larvae pupate and hatch.

- Cover food waste with bedding to prevent another population explosion.

It is important to understand how the fruit fly *Drosophila melanogaster* develops in order to trap it. This insect goes through four stages in its life cycle. An adult fruit fly can live for over 10 weeks and the female can start producing 50 to 70 eggs per day within a week of emerging from the pupal stage. Eggs are laid in fruit (thus the name). Once hatched, the eggs become larvae, which not only feed on the fruit but cause it to rot faster. The larvae molt three times, each time turning into a larger wormlike form. We commonly call this larva a maggot. When the third larva emerges it looks for a clean dry place to begin its pupal stage. Once conditions are good, the pupa becomes a fly and the process begins again.

There are many nontoxic traps commercially available to trap fruit flies. It's also very simple to make a homemade trap using common household items. Fruit flies are attracted to rotting and fermenting fruit. Put apple cider vinegar in a mason jar and add

Homemade traps made of vinegar and liquid soap are effective at catching fruit flies. Pieces of fruit may also be added.

a few drops of liquid soap. The soap breaks the surface tension of the liquid causing the flies to fall in and drown. Don't put a lot of soap in or the smell can overpower the scent of vinegar, which is the smell that pulls them into the jar. Some people suggest adding pieces of rotting fruit to increase the smell. I would only do this if they do not, for some reason, become trapped in the vinegar. After all, eggs are laid in fruit and need it to eat in order to emerge as maggots. Once you have made your jar trap you can cover it with plastic wrap and poke a few holes in the wrap with a toothpick. This allows an easy way for flies to get in but not an easy way for them to fly out of the jar.

Biological controls may work. People have reported some success adding beneficial nematodes to their worm bin. Available from some garden supply catalogs, these nematodes drill into and consume fly larvae and pupae in the worm bin. Used in conjunction with a trap and attractant to reduce the adult fly population, application of beneficial nematodes to the bin may do the trick.

One suggestion came from K. P. Plater, who has used worms to eat his food waste for 45 years. He came upon this possible solution to the fruit fly problem after tens of thousands of ladybugs invaded Bucks County, Pennsylvania, one summer. The following spring, hundreds of ladybugs were on all windows of his home. He captured a handful and placed them underneath the plastic cover on his worm bin. He was delighted to report that he was soon completely free of fruit flies! Mr. Plater acknowledges that it may have been accidental, and further experimentation may be necessary, but at 88 years old, he says he is too old for further trials. Ladybugs are certainly worth a try! Ladybugs are more accurately called lady beetles. They are part of the order Coleoptera, of which there are over 5,000 species. Some are carnivorous, and it is these predatory animals that help gardeners. Their preferred food is aphids, and they don't eat the plants. They will also eat other harmful insects such as fruit flies, thrips, and mites. It follows logic that the lady beetles placed in a worm bin full of fruit flies ate them up.

DISPOSAL

Although this method deserves a minus from the standpoint of environmental soundness because it requires electricity, if you get desperate, suck up flies with a vacuum cleaner. When I lift the lid on my box and lots of fruit flies zoom up and land on the basement ceiling, my vacuum cleaner just inhales them. Although not a complete control, this method does help to reduce the numbers.

OTHER SUGGESTIONS TO PREVENT FLIES

I recommend adding a cup or so of rock dust to a worm bin perhaps twice a year. This by-product from the gravel and rock-crushing industry is gaining in popularity with those who feel that our soils are severely lacking in trace minerals from excessive agricultural activity. They claim that not only will rock dust provide many trace minerals to make the vermicompost from a worm bin far more nutritious for plants, it will help balance the worm bin environment so that flies will be less of a problem.

Black Soldier Fly Larvae
Hermetia illucens (Linnaeus, 1758)

Another organism that might show up in your bin is the larvae of the black soldier fly. These segmented, whitish worms measuring ½" to 1" are decomposers of organic material. You will not be bothered by the adult fly; unlike other flies, they are not a nuisance. They don't buzz or bite and won't come into your house to feed. The first time I encountered the larvae, I was sorry I didn't have a flock of chickens to feed them to. Like worm workers using red wigglers, many worm workers are also raising these whitish worms to feed to their chickens, fish, ducks, and birds.

Ants
(Family Formicidae)

I have never had a problem with ants in my vermicomposting bin, but then I don't have a problem with ants in my home, either. In milder climates, ants could be a problem for which controls must be sought. Although ants may show up in a worm bin looking for food, they prefer dryer conditions. If they are a constant presence, then you may need to increase the moisture in the bin.

  If you do have ants around your bin, then mopping up ant trails with soapy water can temporarily remove foraging ants from an area. This is especially true if the removal of the ant's scent is done at the entry points.

  A paste of borax and sugar mixed with a little water is an effective control for ants, but harmless to people. Setting up physical barriers to prevent their access is also possible. For example, I would set the legs of my worm bin in coffee cans with mineral oil in the bottom. Some worm workers recommend using soapy water. The ants would get trapped in the oil or soapy water and would not be able to enter the bin. Or I might try dabbing petroleum jelly on a piece of cotton and making a continuous 1" swath around the top of my worm bin. If none of these suggestions work, there are a number of commercially available ant traps that use a borate-based product.

DISEASE ORGANISMS

People sometimes ask, "Can you get viruses, germs, or diseases from your worm bin?" That's not a simple question to answer. I have already discussed the potential for transmitting toxoplasmosis if cats are allowed to use a worm bin as a litter box. The organism for this disease is known to pass intact through the digestive tract of an earthworm.

If your cat is harboring the organism, it can pass into the cat's feces. The more you are exposed to the places the feces are deposited, the more likely the organism could enter your body. Of course, this could happen in the absence of a worm bin, whenever you change the litter box for your cat.

For similar reasons, I discourage people from even considering using worm bins as described in this book for treating human manure. Pathogens (disease-producing organisms) can be transmitted in human manure. Our complex and increasingly expensive wastewater treatment facilities are designed to reduce or eliminate the possibility that these organisms will reach our soils and water supplies. Although high-temperature composting has been shown to be effective in killing pathogens, home vermicomposting systems do not generate the high temperatures characteristic of well-constructed, large-mass compost heaps.

Some research suggests that passage through an earthworm's gut can reduce the number of pathogens present in sewage sludges. It is highly probable that vermicomposting can effectively reduce pathogens under certain conditions. Even so, there is no standard operating procedure in place to ensure that all pathogens are eliminated. More research needs to be done in this area. **Never use human waste in a composting system, unless that system is specifically designed for human waste** and the use of worms would enhance its effectiveness.

One further caution: If you are overly sensitive or allergic to fungi and mold spores, you probably won't be able to have a worm bin in your home. Molds can and do develop in natural succession during the composting process. You may have to reserve your vermicomposting activities for outdoor locations, perhaps with someone else doing the required maintenance. Another possibility would be to keep your bin acidity within a pH range of 6 to 8, outside the optimal range for fungi (pH 4 to 6).

CRITTER SUMMARY

You are likely to find many organisms other than earthworms in your worm bin. In truth, the system won't work if they aren't present. Your worm culture is not a monoculture. Instead, it is a diverse, interdependent community of large and small organisms. No one species can possibly overtake all the other species present. They serve as food for each other, clean up each other's debris, convert materials to forms that others can utilize, and control each other's populations. Soils with a high organic content are likely to contain great numbers of soil organisms. Since the nature of vermicompost is basically organic, no one should be surprised that it also contains soil organisms in great variety and large numbers.

For us to arbitrarily decide who should live and who should die in this complex system is a bit presumptuous. Although some controls are suggested, this chapter's major purpose is to provide a better idea of what you can expect to find. Don't be alarmed at what you see. Information tempers fear. You may even decide to learn more about those critters that you once used to squash when you found them.

HOW TO USE YOUR VERMICOMPOST

Having worms eat your food waste can lead to healthier plants. This happens when you use the vermicompost from your worm bin on your houseplants and gardens. What is the nature of this rich humus, and how should you use it?

Remember the distinction between worm castings and vermicompost. Worm castings are deposits that once moved through the digestive tract of a worm. Vermicompost is a dark mixture of worm castings, organic material, and bedding in varying stages of decomposition, plus the living earthworms, cocoons, and other organisms present.

COMPLETING THE CIRCLE

If you choose a low-maintenance system, a large proportion of your vermicompost will be worm castings. A worm casting (also known as worm cast or vermicast) is a biologically active mass containing thousands of bacteria, enzymes, and remnants of plant materials and animal manures that were not digested by the earthworm. The composting process continues after a worm casting has been deposited. In fact, the bacterial population of a cast is much greater than the bacterial population of either ingested soil or the earthworm's gut.

An important component of vermicompost is humus. Humus is a complex material formed during the breakdown of organic matter. One of its components, humic acid, provides many binding sites for plant nutrients, such as calcium, iron, potassium, sulfur, and phosphorus. These nutrients are stored in the humic acid molecule in a form readily available to plants and are released when the plants require them. Humus increases the aggregation of soil particles, which in turn enhances permeability of the soil to water and air. It also buffers the soil, reducing the detrimental effects of excessively acid or alkaline soils. Additionally, humus has been shown to stimulate plant growth and to exert a beneficial control on plant pathogens, harmful fungi, plant parasitic nematodes, and harmful bacteria. One of the basic tenets of gardening organically is to carry out procedures that increase the humus component of the soil; earthworm activity certainly does this.

WHERE TO USE VERMICOMPOST

You will have several buckets full of vermicompost from your worm bin. Use it selectively and sparingly. Vermicompost is loaded with humus, worm castings, and decomposing matter. The cocoons and worms present are unlikely to survive long outside the comfort of your bin. Plant nutrients will be present, both in stored and immediately available forms. Vermicompost in sufficient quantities also helps to hold moisture in the soil, which is an added advantage during dry periods.

Seed Beds

Vermicompost will not "burn" your plants as some commercial fertilizers do, but since your supply will be limited, use it only where it will do the most good. One method is to prepare your seed row with a hoe, making a shallow, narrow trench. Sprinkle vermicompost into the seed row. In this way, the new seeds will have the vermicompost as a rich source of nutrients soon after they germinate and during early stages of their growth.

Transplants

For transplanting such favorites as cabbage, broccoli, and tomatoes, which are usually set out in the garden as young plants, throw a handful of vermicompost in the bottom of each planting hole you dig. Don't worry if worms or cocoons are present in the vermicompost. While the worms are alive, they will produce castings and add nitrogen from their mucus, but they are not likely to do all the other good things that worms do for the soil. Also,

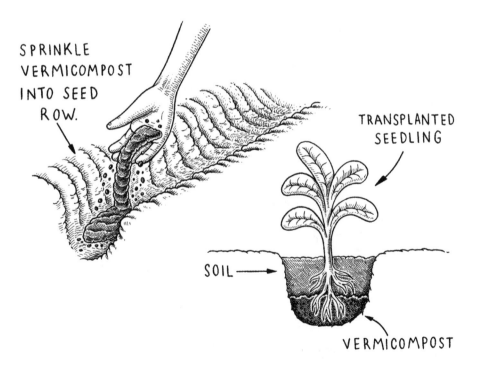

SPRINKLE VERMICOMPOST INTO SEED ROW.

TRANSPLANTED SEEDLING

SOIL

VERMICOMPOST

don't expect your redworms to thrive in your garden. They are not normally a soil-dwelling worm, and they require large amounts of organic material to live. If you were to add large quantities of manure, leaves, or other organic material, you might find that a few *Eisenia fetida* survive, but most will probably die. When they do, their bodies will add needed nitrogen to the soil, so all is not lost! Hopefully, your gardening techniques will improve the organic matter concentrations in your garden so that the soil-dwelling species of earthworms will be fruitful and multiply.

Topdressing

You will use most of your winter production of vermicompost during spring planting. Any remaining material can be applied later in the season as a topdressing or side-dressing. At this time, you won't want to disturb the growing plants' root systems, but it is a simple matter to sprinkle vermicompost around the base and dripline of your plants, giving them an additional supply of nutrients, providing organic matter, and enabling the midseason plants to benefit from vermicompost's water-holding capacity.

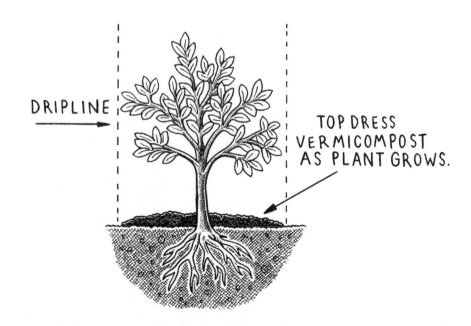

DRIPLINE

TOP DRESS VERMICOMPOST AS PLANT GROWS.

HOW TO USE WORM CASTINGS

After several months of low-maintenance technique, the contents of the worm bin will be a dark, crumbly material that smells like earth. Since little food is left in this material for earthworms, very few worms will be present. Populations of active microorganisms will also have dwindled; those remaining will be in a dormant state awaiting reactivation in a suitable environment of new food and moisture.

Except for some large chunks, most of this worm-bin material is worm castings. Worm castings differ from vermicompost in being more homogeneous, with few pieces of recognizable bedding or food waste. When dried and screened, castings look so much like plain, black organic topsoil that you may be surprised to recall that little or no soil went into the original bedding.

While some drying of worm castings is desirable, it is best not to let them dry to the point that they become powdery, for it then becomes difficult to wet them down. A crust may form on the surface, causing slow water penetration. Worm castings with about 25 to 35 percent moisture have a good, crumbly texture and earthy smell, and are just about right to use on your plants.

Chemistry and Castings

Although pure worm castings provide many nutrients for plants in a form the plants can use as needed, some precautions should be taken in applying the castings. The organic material present in food waste likely will have broken down to a greater extent in worm castings than in vermicompost. More carbon will have been oxidized and given off as carbon dioxide, leaving nitrogen, phosphorus, potassium, calcium, and other elements to combine to form various salts.

Worm castings may not serve as a "complete" fertilizer for some plants. For one thing, high concentrations of some salts can inhibit plant growth. Additionally, worm castings often have a pH of 8 or higher and thus may not be suitable for acid-loving plants. The solution is to mix worm castings with other potting materials. In this way, plants gain the advantage of the nutrients present without suffering from excessive salt concentrations or changes in pH (see page 13 for information on pH).

An intriguing experiment that seemed to verify the need to dilute pure castings was conducted by a horticulturist at the Kalamazoo Nature Center in the 1970s. Three sets of African violet plants were potted, each set in a different medium. Some grew in 100 percent potting soil; some in 100 percent worm castings; and some in equal amounts of worm castings, perlite, and Michigan peat. When the plants grown in potting soil were compared with those grown in worm castings, the castings-grown plants were generally healthier. Those grown in the potting soil showed chlorosis (yellowing) of some leaves, a sign of possible nutrient deficiency.

The plants grown with worm castings diluted with perlite and peat were distinctly more vigorous than the other two sets of plants. Leaves were larger, greener, and more robust. A likely interpretation of this experiment is that although the pure castings provided more nutrients for the young plants than the potting soil, salt concentrations in the castings may have been great enough to inhibit their growth. The plants in the third set had the benefit of nutrients from the castings but were not inhibited by too high a concentration of salts, since the concentration had been reduced by dilution with the perlite and peat.

In the 1980s, Dr. Clive Edwards, now of Ohio State University, led a team of scientists who conducted extensive plant growth studies that compared worm castings to commercial potting media in his native England. Castings were produced by worms working on a variety of wastes, such as cattle and pig manures, brewery waste, and potato waste. Although adding magnesium and adjusting pH to decrease alkalinity were sometimes necessary, media containing worm castings produced plants with as good or better seed germination and plant growth, and earlier flowering. This was true even when worm castings comprised as little as 5 percent of the mixture. Other more recent work supports these findings. It should be noted that reviews of this study use the terms *worm castings* and *vermicompost* interchangeably.

A 2007 study by Edwards and Dr. Norman Arancon confirmed that vermicompost can increase germination, growth, and yields of crops. Flowers, vegetables, and fruits were grown in greenhouses and fields. The researchers found that growth rates and yields improved more with lower applications than with higher ones. A 100 percent vermicompost application yielded less than a 20 to 40 percent vermicompost mixture. Again, it is possible that this was in part due to higher amounts of inorganic salts in the compost. This study also showed that large amounts of plant growth hormones exist in vermicompost. The researchers concluded that a wide range of plant pests were suppressed in vermicompost and that plant pathogens were reduced, as were plant parasitic nematodes.

Investigations are continually under way to refine this knowledge further. As scientific and commercial development of vermicomposting proceeds, the true economic value of worm castings will become apparent. In the meantime, your plants certainly will benefit from the castings the worms in your worm bin produce.

To Sterilize or Not to Sterilize

Some people suggest "sterilizing" potting mixes and worm castings prior to use with houseplants and in greenhouses in order to kill organisms that could cause plants trouble in a confined environment. The term "sterilizing" is being used loosely here, since the dictionary definition of sterilization is "the destruction of all living microorganisms, as pathogenic or saprophytic bacteria, vegetative forms, and spores." Surgical instruments, for example, are sterilized in an autoclave under high temperature and pressure for a specified period of time. For our purposes, it would be more correct to say that potting soil is pasteurized; that is, it is exposed to a high temperature or poisonous gas for a long enough period of time to kill certain microorganisms, but not all.

In any case, whether you prefer to call it sterilize or pasteurize, I don't recommend that you do either to worm castings. Soil (vermicompost) is a dynamic, living entity, and much of its value comes from the millions of microorganisms present.

One concern many people voice about using worm castings directly on their houseplants is "Won't those little white worms and all those bugs I can see crawling around hurt my plants?" Probably not. The enchytraeids eat dead and decaying material, not living plants, and so do the mites and springtails that are likely to still be present when your vermicompost is almost all worm castings. The organisms that thrived in your worm box are not likely to be the kind that also attack living plants. If there are just a few, don't worry about them.

There may be a lot. If you have a true aversion to having visible critters in the worm castings you want to sprinkle under

An Added Bonus

Have you ever tried to germinate an avocado pit? Have you tried the trick with the three toothpicks, inserting them around the diameter of the pit, placing it on top of a jar of water, and keeping it watered for . . . well, months? Until you either got tired of it or it finally did sprout?

Well, have I got a deal for you! Throw your pits into your worm bin, cover, and forget about them. That's all. In time — it may take months, but be patient — you will find a taproot coming out of the bottom and a sprout coming out of the top. When this happens, transfer it to a pot. One winter, nine out of ten avocado pits I tried this way germinated. I now have more avocado plants than I know what to do with — in the living room, on the front porch, on the side porch, in my office. . . You, too, can be the first on your block to be a success with your sprouted avocado pits.

your plants, place your worm castings on a sheet of plastic out-doors in the sun. Put another sheet of plastic on top, and let this "solar heater" warm things up a bit. Most of the white worms will move onto the plastic, and most of the mites and springtails will be killed from the heat. Collect your castings in a few hours. They will be ready to use in potting mixes, as topdressing, or in your garden.

Potting Mixes

Worm castings may be mixed with various concentrations of potting materials, such as coir, sand, topsoil, perlite, vermiculite, or leaf mold. One satisfactory mix of equal parts by volume has these ingredients:

- ¼ worm castings: for nutrients
- ¼ coir: for moisture retention
- ¼ perlite: for aeration
- ¼ sand or garden soil: for body

Experiment with different mixes, and find the ones that suit your favorite plants.

Soil Blocks

A device called a soil blocker can make a few worm castings go a long way. This small, hand-operated plastic or metal tool enables you to compress a potting soil mixture into various sizes. Block forms can range from ¾" (1.9 cm) to 4" (10.2 cm). The size you choose will depend on the seed you are planting and the length of time you intend to grow the seed before transplant-ing. The top of the block will have either an indentation to drop a seed into or a deep center hole for inserting a root cutting or transplant. Make up a wetter than normal potting mix containing about 25 percent worm castings, and compress it into the soil blocker. Plant a seed in the resulting block; keep it moist, as you would a peat pot. When the young plant is ready to transplant, insert the soil block into its planting hole and the young plant will take off without undergoing the normal transplant shock. When a plant is started in a traditional container, its roots will circle around the container wall, and transplant shock can occur when the seedling goes into the ground. In contrast, soil blocks are placed on trays with air space between the block walls. Shock is reduced because the plant roots grow to the edge of the soil wall and wait to grow further until they are either put into a larger soil block or into the ground. Another advantage of starting seeds this way is the environmental savings of not using a plastic pot, especially those that are used once and thrown away into a landfill. And as already mentioned, I do not recommend the use of peat.

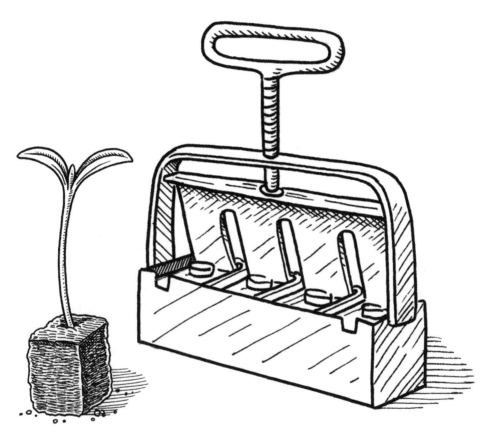

Using a soil blocker with a mix of potting soil and worm castings is efficient and prevents transplant shock.

Topdressing

Sprinkle a layer of worm castings about 1/4" deep on the soil surface that supports your potted plants, and water as usual. Repeat
every 45 to 60 days. If necessary, remove some of the soil above
the roots so that you have room for the worm castings. Let
excess water move through the soil occasionally to flush out
accumulations of salts, particularly if you have hard water. And
remember, don't use softened water on your plants; it contains
other salts that do harm.

In your garden, sprinkle worm castings along the bottom of
your seed row or throw a handful of castings into the hole when
you are transplanting. The adjacent soil will dilute excessive salt
concentrations in the castings. It is perfectly natural for vegetable seeds present in the vermicompost to sprout. Simply pull
these sprouts as you would pull weeds.

TREATING WASTE AS A RESOURCE

The process of recycling returns materials to a previous stage in a repetitive process. In a natural ecosystem, the wastes from one process become the resources for other processes, a concept discussed earlier in chapter 11.

In other words, there is no waste in nature, but a continuous flow of materials and energy from one organism to another through basic cycles such as the carbon, nitrogen, and water cycle. Nature uses a circular system. Our present Western society uses a linear system: virgin materials are extracted, processed via manufacturing, and consumed by buyers, who then throw away the waste. Each of these four steps uses energy and contributes to pollution.

When the second edition of this book was released in 1997, some initiatives had begun changing our thinking from a linear system to a circular one. Canberra, Australia's capital city, was the world's first community to establish the goal "No waste by 2010." For its citizens and governmental leaders, a waste-free society is one in which no material is regarded as useless, where all resources find another application or useful function. Meanwhile in Berkeley, California, an organization and building materials exchange called Urban Ore was beginning to work toward its goal of total recycling. Although Canberra's initiative did not reach the zero waste goal due in part to a change in governmental direction, it did help establish zero waste as a goal throughout the world. Although zero waste is not yet the norm, it is encouraging that waste reduction and recycling programs are now widespread and successful in many places around the world.

ZERO WASTE

There are a number of definitions of zero waste. The one I use was developed by the Zero Waste International Alliance (ZWIA). It reads: "Zero Waste is a goal that is ethical, economical, efficient and visionary, to guide people in changing their lifestyles and practices to emulate sustainable natural cycles, where all discarded materials are designed to become resources for others to use. Zero Waste means designing and managing products and processes to systematically avoid and eliminate the volume and toxicity of waste and materials, conserve and recover all resources and not burn or bury them. Implementing Zero Waste will eliminate all discharges to land, water or air that are a threat to planetary, human, animal or plant health."

Most zero waste goals are set at a 90 percent diversion rate from landfills, incinerators, and the environment. So it actually is not a 100 percent goal. Nonetheless it is an ambitious and obtainable target.

We have all heard of the bottom line, but have you heard of the triple bottom line? Abbreviated *3BL*, it's also known as the three pillars or the three Ps: profit, people, and the planet. More companies are adopting the Zero Waste Business Principles developed by ZWIA, principles that encourage looking at resources in new ways. Companies are finding that it benefits their profit margins to consider employees and the planet. For more information on these business principles, see Resources (page 176).

Zero Waste, Worms, and Food Waste

Worm bins could contribute significantly to achieving the goal of a waste-free or zero waste society. Food waste turns into plant nutrients with a little help from the worms.

A big advantage of worm bins is that this recycling takes place on-site, close to (if not in) people's homes. People are willing to have worm bins because they can eliminate smelly kitchen wastebaskets. Trash containers will be less likely to attract flies; lifting their lids won't cause people to reel from the fumes. I've often said, "The time to publicize a worm-composting program is three weeks into a garbage haulers' strike in a city!"

Food waste now accounts for 18 percent of the waste stream in the United States. That is more than 30 million tons of food wasted each year. Keeping food out of the landfill saves limited landfill space, but it also keeps that food from producing methane, which is a potent greenhouse gas. Additionally, organic materials lead to the production of a leachate. As these liquids seep through the landfill, they pick up toxic chemicals and metals. If they leak through the landfill liner, they can potentially contaminate groundwater. In 2010, the U.S. Environmental Protection Agency started its Food Recovery Challenge to divert food from landfills by prevention, donation, and composting. This strategy follows the three pillars: prevention increases profits, donations help people, and composting benefits the planet. Today there are more than 800 businesses and organizations that participate in or endorse this challenge (see Resources, page 176).

Fortunately, many more citizens have access to recycling programs than they did in the early 1980s, when the first edition of this book came out. At that time, only about 700 communities had curbside recycling programs. In 1996, citizens in nearly 9,000 cities could participate in recycling just by taking their recyclables to the curb. San Francisco was the first U.S. city to mandate composting. Today there are 90 cities that have taken this green mind-set one step further with curbside *composting*. These numbers are encouraging, but given that many U.S. cities still don't have curbside recycling, let alone curbside composting, it makes it even more important to compost your food waste.

With a worm bin in your home, you will find it more convenient to examine some of the materials you throw away. Think about where they go. Can more of those materials be recycled? Aren't we going to run out of raw materials if we just keep throwing things away after one use? Can you come closer to having a zero waste household?

Why has the public embraced recycling wholeheartedly? Perhaps it's because people sense the practicality of the old adage "Waste not, want not." Recycling saves materials by turning them back into valuable products rather than burning them in an incinerator or burying them in a landfill. And recycling saves energy. Recycling the aluminum in a soda can takes about 6 percent of the energy it takes to produce that aluminum from bauxite.

Obviously, we have a long way to go to achieve a zero waste society. Fortunately, there are some things you can do now.

THE THREE Rs AND BEYOND

For years people have talked about the three Rs; reduce, reuse, and recycle. Today it is more like the five or six — or more — Rs. Here are some of these important concepts.

Rethink

An advantage of having your own worm bin is that instead of thinking of food scraps as waste, you think of them as a resource. This can spur a paradigm shift: you might start looking at other things as resources instead of as waste products. We need to take responsibility for our impact on the planet. It is easy to assume that governments or industries are going to take care of the issues that keep us from living in a sustainable system. But we as individuals need to make choices and be knowledgeable to influence what our governments and corporations decide.

Reduce and Refuse

Reduce the amount of materials that flow in and out of your home. Buy for quality and longevity when possible. Refuse excess packaging by buying items that have less packaging. Refuse that plastic bag in the store. Bring your own reusable bag to cart items home. Not all restaurants use recyclable, reusable, or compostable take-home containers — which is why I keep extra containers in my car for that purpose. Each of our individual actions is important, as are those that influence cities to take action and change our choices.

More and more states and cities are passing legislative bills regarding the regulation and banning of retail plastic bags. We know that these bags choke wildlife, don't decompose, become litter easily, and add to our demand for oil. Some stores take bags back to recycle; however, it is estimated that in the United States nearly 90 percent of these bags are not recycled. Some places, such as California, have implemented broad, statewide bans on single-use plastic bags. Although it will take time to study and fully understand the impact of bills such as these, we have some promising early examples to draw from: over a four-year period after banning plastic bags, the city of San Jose, California, found a 78-percent reduction of plastic bags in their creeks and rivers and a 69-percent reduction in storm drain inlets.

American society does not make it easy to reduce materials consumption. It is difficult to determine the economic cost of recycling versus landfills. As a retired CPA, I am aware of the old joke "What is two plus two?" — with the answer being "What would you like it to be?" The cost of using virgin materials drops as oil prices drop. A true cost equation would look at the triple bottom line. It would factor in the costs to humans and to the planet and not just a short-term profit margin. In the past 20 years, recycling has become an issue for large corporations. For some it is easier to fund large recycling facilities then to reduce their dependence on plastic. There are many players in the future of our planet, so be sure to be informed. Use the National Conference of State Legislature (NCSL) website (see Resources, page 176) to learn which states are trying to ban one-use bags and which are trying to ban local efforts to ban those bags.

Reuse

Reuse what you can. Buy a reusable water bottle. Shop for used items. Donate what you don't want or need so others can have them. There are many opportunities to pass on items we no longer need. A number of nonprofit corporations operate stores that encourage reuse of items from clothing to building materials. Some of these retail centers — like Salvation Army, Goodwill, and Habitat for Humanity ReStore — can be found in nearly every state. Others might be neighborhood shops that help fund a local nonprofit. Another way to donate is to check around your area for churches, schools, businesses, and community centers that organize collections. In my area, food banks and organizations feed the hungry. During winter, our local electric company collects coats and other cold-weather gear to donate to people in need. Little Free Libraries are popping up everywhere as a way to recycle books. Fashioned after this idea, one of our local churches set up the pilot program for a Little Free Pantry where people can drop off necessities such as food and household supplies for others to pick up. The Medical Supply Network, originally started by the Rotary Club of Tulsa, collects medical supplies and ships them around the world.

Develop a conservation mentality (some people might say a "Depression mind-set," referring to the frugal tenor of the 1930s) about the use of materials. What you think about what you do determines your attitude. Attitude, of course, is a little thing that makes a big difference. Thinking "stewardship" feels better than thinking "deprivation." The better you feel about your behaviors, the longer you'll be willing to maintain them. People have shared with me that composting their food waste with worms makes them feel good. They like turning the most repulsive part of their trash into something useful. Let companies know what you think. Encourage production of packaging that can be reused and recycled, rather than mixtures of plastics and paper that are not good for either reuse or recycling.

Repair

Maintaining and repairing your belongings saves much more energy and resources than replacing them when they break. Fixing things has become easier with the advent of online videos and how-tos. I personally fixed my dryer by looking online for help. The Center for EcoTechnology has links to fixing a variety of electronics (see Resources, page 176). After I finish this chapter, I am going to see if I can repair my old laptop!

Rot

You can compost with or without worms. I do both. New York City offers residents compost bins at cost to encourage urban composting, and they also have a broad educational system to support composting. See if your town or city offers similar programs.

Recycle

Recycle all of your plastics, newspapers, cans, glass, and aluminum. As the cost of obtaining virgin materials increases, incentives for recycling already processed materials also increase, creating larger and more reliable markets for recycled materials.

In the United States, landfill fees are relatively low, and our waste-management system is fragmented. As of this writing, this country's recycling rate is about 34 percent. As mentioned previously, it takes 94 percent less energy to make cans from recycled aluminum, yet 40 billion cans end up in landfills annually. What makes this figure even more interesting is that we have container-deposit laws in ten states, and these states account for 48 percent of the recycled containers. If a state wants to increase the amount of recycled containers, it seems fairly apparent that container-deposit laws work. To learn which states have laws and how those laws read, you can visit to the NCSL website (see Resources, page 176).

Plastic poses many issues, to say the least. Americans use 4 million plastic bottles each hour, and only one in four will be recycled. Nearly 8 million tons of plastic go into the ocean annually, which is causing immeasurable damage to marine life. Fish, marine mammals, and seabirds choke on the colorful plastic as they mistake it for food or accidently ingest tiny pieces of plastic circulating in the polluted water. Unfortunately, that plastic litter in the ocean is estimated to increase to 80 million tons per year by 2025. By 2050, it is estimated that the weight of plastic in the ocean will be equal to the weight of fish.

Both economic and political forces influence the direction of municipal waste programs. You can help by striving for environmentally sound programs not only in your own home but in your community. For example, there is a push to make it easier on consumers by not requiring sorting of materials, but opponents argue that mixing materials causes contamination and therefore reduces the value of the recycled material.

Differential rates for trash pickup are becoming more common, creating a financial incentive to produce less waste. Many cities now charge for trash collection based on the size of the container. Under such systems, every family who normally uses the largest container each week could cut back to the smallest by installing a home vermicomposting system and recycling all recyclables. This possibility may not exist in your area because of garbage-collection practices. However, money could be saved and made available for other things when municipalities and waste-disposal firms do the right thing. Landfills will last longer. Toxic leachates are less likely to develop, since there will be less organic acid from the food waste to react with metals and other materials in the dump. Libraries, community centers, parks, and recreation services could be supported with money that is currently used for landfills. Families could pay less for garden fertilizer, less to purchase fishing bait, and . . . and . . . and . . . why isn't everybody doing it?

Appendix A

Record Sheet

Date set up _____ = Day 0

Description of setup: _____

Initial weight of worms: _____

☐ Breeders or ☐ mixed sizes

Type of bedding: _____

Size of bin: _____

Number in household: _____

Food waste burying locations:

	2'				3'			
	1	6	7		1	6	7	12
2'	2	5	8	2'	2	5	8	11
	3	4	9		3	4	9	10

Date	Day	# oz.	Total # oz. to date	Temp.	Water # of pints	Burying loca-tion #	Comments

Date Harvested: _____

No. of Days: _____

Worm weight:_____

Total weight garbage buried:_____oz. = _____lb.

Weight uneaten garbage: _____

Average oz. buried per day: _____

Average temp.: _____

Temp. range:_____

Appendix B
Annotated References

Earthworm Books for the Layperson

Edwards, Ray. *The Nightcrawler Manual*, 5th ed. Eagle River, WI: Shields
    Publications, 1990, 1976.

This manual is one of few that gives information about keeping — not breeding —
nightcrawlers. Edwards, who has studied nightcrawlers for years, tells how to
harvest, hold, feed, water, and condition them for fishing. Breeding nightcrawlers
(*Lumbricus terrestris*) is not practical because of their burrowing behavior and
requirement for large volumes of soil to inhabit.

Ernst, David. *The Farmer's Earthworm Handbook: Managing Your Underground
    Money-makers*. Brookfield, WI: Lessiter Publications, 1995.

David Ernst gives us a practical, well-researched book written by an obvious worm
enthusiast. He goes to the scientific literature for documentation on the effects of
earthworms on agricultural crops. Farmers share their experiences switching from
conventional tillage practices (which reduce earthworm populations) to those
that allow the earthworm populations to come back. Ernst presents information
on identification of common earthworkers, tillage practices, manure management,
chemical effects on earthworms, and cover crops.

Hale, Cindy. *Earthworms of the Great Lakes*. Duluth, MN: Kollath + Stensaas
    Publishing, 2007, 2013.

An important book explaining issues regarding the earthworm as an invasive and
what we can do about it. Hale includes information on how to safely compost with
worms. There are sections on how to identify various species with color photographs and an earthworm key. The book also explains how to collect and preserve
earthworms and, if interested, how to add specimens to the Great Lakes Worm
Watch archive.

Hopp, Henry. *What Every Gardener Should Know about Earthworms*. Pownal,
    VT: Storey Communications, 1978.

Although much of the information in this meaty little booklet was adapted from a
1954 publication entitled "Let an Earthworm Be Your Garbage Man," the presentation on effects of earthworms on soil moisture, aeration, and soil fertility is still
pertinent.

Stewart, Amy. *The Earth Moved: On the Remarkable Achievements of
    Earthworms*. Chapel Hill, NC: Algonquin Books of Chapel Hill, 2004.

It is unusual for an author to take a subject like earthworms and produce a book
that the reader only puts down because he or she needs to do something else.
Stewart has produced such a book. Written by a scholar with a gardener's view-
point, this book contains a wealth of information. The important facts are covered,
plus information and interviews with earthworm specialists. Although the main subject of the book is earthworms, I found many new ideas for gardening in this book.

Earthworm Books for the Advanced Student

Edwards, Clive A., Norman Q. Arancon, and Rhonda L. Sherman
(eds). *Vermiculture Technology, Earthworms, Organic Wastes and Environmental Management*. Boca Raton, FL: CRC Press, 2011.

In 2000, Mary Appelhof collaborated with Edwards to organize the Vermillenium. Proceedings from that conference were not published until now. This book contains those updated proceedings and more, resulting in a review of all aspects of the science of vermiculture technology.

Earthworm Farming

Barrett, Thomas J. *Harnessing the Earthworm*. Boston: Wedgewood Press, 1947, 1959.

This important document synthesizes much of the early literature on the effects of earthworms on soil fertility. Discusses humus, topsoil, subsoil, earthworm tillage, and chemical composition of earthworm castings. Excellent and still useful information on earthworm culture is provided.

Bogdanov, Peter. *Commercial Vermiculture: How to Build a Thriving Business in Redworms*. Prescott Valley, AZ: Petros Publishing Co., 1996.

With its focus on how to make money raising earthworms, this book is a welcome source of information on the business of vermiculture. Bogdanov puts vermiculture into a historical context, gives basic information about composting worms, tells how to get started, and describes how to set up commercial beds. He covers pests and predators, harvesting, and packaging and shipping. A must for anyone wanting to go into the worm business.

Brown, Amy. *Earthworms Unlimited*. Kenthurst, NSW, Australia: Kangaroo Press Pty., 1994.

A long-standing fascination with worms and the idea of setting up a potential source of income for a relative stimulated Brown to collect 1,400 worms from a compost pile. She used information from the basic worm growers' manuals and her own experience to sort out types of bins, bedding, feeds, procedures, packaging, and marketing aspects of worm growing. She convinced herself that a viable business could be developed, but she sold her experimental worm farm so that she could go back to her full-time profession of writing. She shares the knowledge she gained in a humorous, informative manner and provides in one little book what would take several other books to uncover.

Holcombe, Dan, and John J. Longfellow. *OSCR: Blueprint for a Successful Vermiculture Compost System*, Portland, OR: Oregon Soil Corporation, 1995, 2003.

This manual provides construction details and operation procedures for a self-harvesting vermicomposting unit made of wood. Plans include directions to make a harvesting screen. The OSCR system is based upon the successful large-scale vermicomposting reactor Holcombe operates to process several tons of organic waste per day.

Shields, Earl B. *Raising Earthworms for Profit*, 20th ed. Eagle River, WI: Shields Publications, 1994, 1959.

This manual has been the standard training manual for hundreds, if not thousands, of worm growers. It discusses markets, propagation boxes, indoor and outdoor pits, feeds, packing and shipping, and advertising. The basic text was written in the 1950s.

Composting

Martin, Deborah L., and Grace Gershuny (eds.) *The Rodale Book of Composting*. Emmaus, PA: Rodale Press, 1992.

A comprehensive, readable book that gives history, benefits, techniques, materials, and machines related to composting. For those who want to know everything about composting.

Soil Animals

Lavies, Bianca. *Compost Critters*. New York: Dutton Children's Books, 1993.

A superb collection of color photographs of the smaller denizens of a compost pile such as springtails, sow bugs, and mites. The text describes the sequence of events and creatures that gradually change organic garbage from weeds, leaves, fruit, and vegetable leftovers into moist, dark, nutritious material perfect for plants.

Worm Books for Kids

Glaser, Linda. *Wonderful Worms*. Brookfield, CT: Millbrook Press, 1992, 1994.

Illustrations provide unusual perspectives for showing worms. A cross-section of soil shows feet walking on it, plants growing on top with roots underneath, or rocks nestled into the earth.

Henwood, Chris. *Keeping Minibeasts: Earthworms*. New York: Franklin Watts, 1991.

The outstanding color photographs of earthworms in this book geared for young children make it interesting for adults as well. The book guides a child in finding, collecting, and caring for one or more soil-dwelling earthworms.

Himmelman, John. *An Earthworm's Life*. New York: Children's Press, 2001.

A beautifully illustrated book for ages five and up. From the Nature Upclose Series, the book contains interesting information about worms and would be useful for teaching about the soil and the environment. I appreciated that although the book is geared toward elementary-school children, the author presents the information in a format that introduces the reader to more complex ideas.

Ross, Michael Elsohn. *Wormology*. Minneapolis: Carolrhoda Books, 1948.

A charming book with much kid appeal. It supports and encourages learning by inquiry by garnering questions from kids in second through sixth grades and devising ways to get earthworms to answer their questions. Part of the Backyard Buddies series.

Metric Conversions

Length

To convert	to	multiply
inches	millimeters	inches by 25.4
inches	centimeters	inches by 2.54
inches	meters	inches by 0.0254
feet	centimeters	feet by 30.48
feet	meters	feet by 0.3048
yards	meters	yards by 0.9144

Volume

To convert	to	multiply
ounces	grams	ounces by 28.35
pounds	grams	pounds by 453.5
pounds	kilograms	pounds by 0.45

Temperature

To convert	to	
Fahrenheit	Celsius	subtract 32 from Fahrenheit temperature, multiply by 5, then divide by 9

Appendix C
How Many Worms in an Acre or a Hectare?*

*Metric conversions for these quantities and areas are approximate.

One year I counted and weighed all the earthworms I could hand sort from the top 7" of a square foot dug from my garden's soil. I counted 62 worms of all sizes, and found at least two species. If I had had an acre under cultivation and if this were, in fact, a representative sample, the total population would have exceeded 2.7 million worms per acre — or 6.8 million worms per hectare!

In U.S. measurement units, these 62 worms weighed 2 ounces. Extended to 1 acre, this would give a total weight of 5,445 pounds, or over 2.5 tons of worms in the top 7" of 1 acre of soil. In metric units, the 62 worms weighed 56.7 grams. Extended to 1 hectare, this would give a total weight of 6,113 kilograms, or more than 6 metric tons of worms in the top 17.8 centimeters of 1 hectare (10,000 square meters) of soil.

How Much Do Their Castings Weigh?

Earthworms of the soil-dwelling type eat soil in their search for organic nutrients. They mix the soil with organic materials and bacteria in their intestines, and excrete the mixed deposit as castings. The weight of these castings per worm per day could easily equal the weight of the worm. Let's estimate conservatively that the weight of castings deposited per day from one worm is one-eighth the worm's weight. The total weight of castings produced per acre per day would be 680 pounds (308.4 kg). Think of the worm's activity in producing those castings and their value to plants.

To estimate annual casting production, let's assume that the worms are active only 150 days of the year, giving 102,093 pounds per acre per year, or over 51 tons of castings per year; that's 112 metric tons of castings per hectare. (If you have ever tossed a ton of manure onto a pickup truck and off again, you can begin to appreciate the work worms do for you in your garden.) Charles Darwin's approach in 1881 to estimating castings production was to collect, dry, and weigh all castings from a square yard of grass for a whole year. He recorded from 3 to 16 tons of dry earth annually ejected by worms in the form of castings.

How Do Agricultural Practices Affect Earthworms?

Some modern agricultural practices have reduced not only earthworm and microbial populations but also the amounts of organic matter present in the soil. Less organic matter results in less food for worms. In addition, plowing and tilling not only kills worms directly, but it exposes them to predators and drying conditions. Worm burrows provide a means for water to move deeper into the soil. Plowing destroys worm burrows, reducing the capacity of the soil to hold water and increasing the possibility of flooding during torrential rains.

Crop management practices that result in increased worm populations have also demonstrated higher crop yields. Such practices include less tillage, planting cover crops, and retaining stubble, all methods that retain moisture and provide food for the worms. An example reported by John Buckerfield from Australia showed substantial grape yield increases when straw mulch and composted yard trimmings were applied underneath the vines. The increase in crop yield correlated with increased earthworm populations. Other promising research showed that applying vermicomposted grape residue and cattle manure to vineyards and other horticultural crops increased yields. Applied under mulch, these applications also reduced the need for irrigation and weed control.

Apparently, many climatic regions could benefit from applying this knowledge related to earthworms, whether they be earthworkers or vermicomposters. Part of my work is to help people understand how earthworms enhance the quality of our soils, our food, and our lives. I also want to discourage those practices that kill these amazing creatures. Their production of natural manure has been creating and improving soils for millions of years.

Glossary

acid: A normal product of decomposition. Redworms do best in a slightly acidic (pH just less than 7) environment. A pH below 5 can be toxic. Addition of pulverized eggshells and/or lime helps to neutralize acids in a worm bin. *See also* pH.

ACQ: Alkaline Copper Quaternary. A wood preservative made by dissolving copper in a solution. Used as an alternative to CCA.

aggregation: The clustering, as of soil particles, to form granules that aid in aeration and water penetration.

aeration: Exposure of a medium to air to allow exchange of gases.

aerobic: Pertaining to the presence of free oxygen, or to organisms that utilize oxygen to carry out life functions.

albumin: A protein in cocoons that serves as a food source for embryonic worms.

alkaline: Containing bases (hydroxides, carbonates) that neutralize acids to form salts. *See also* pH.

anaerobic: Pertaining to the absence of free oxygen, or to organisms that can grow without oxygen present.

anaerobiosis: Life in an environment without oxygen or air.

anterior: Toward the front.

Ardox nails: Nails with a spiral shank designed to increase holding power.

bedding: A moisture-retaining medium used to house worms.

bed run: Worms of all sizes, as contrasted with selected breeders. Also known as pit-run, run-of-pit.

biodegradable: Capable of being broken down into simpler components by living organisms.

biological control: The management of pests within reasonable limits by encouraging natural predator/prey relationships and avoiding use of toxic chemicals.

biomass: That part of a given habitat consisting of living matter, expressed as weight of organisms per unit area. The recommended biomass of worms for vermicomposting is about 1 pound per square foot (0.5 kilogram per 0.1 square meter) surface area of bedding.

breeders: Sexually mature worms as identified by a clitellum.

buffer: A substance that renders a system less sensitive to fluctuations between acidity and alkalinity. Humus serves as a buffer in soil.

CA: Copper azole. A wood preservative made by dissolving copper in a solution. Used as an alternative to CCA.

calcium carbonate: Used to reduce acidity in worm bins and agricultural soils. *See also* lime.

castings: *See* worm castings; vermicast.

CCA: A term for wood treated with a preservative containing copper, chromium, and arsenic designed to control against termite and fungus damage.

CDX plywood: CD plywood has knotholes and small splits present (in contrast to a higher grade, such as AB, which has one side smooth and free from defects). Exterior (X) plywood is bonded with waterproof glue and suitable for use outside.

cellulose: An inert compound containing carbon, hydrogen, and oxygen that is a component of worm beddings. Wood, cotton, hemp, and paper fibers are primarily cellulose.

chlorosis: The abnormal yellowing of plant tissues caused by nutrient deficiency or activities of a pathogen.

clitellum: A swollen region containing gland cells that secrete the cocoon material. Also called girdle or saddle.

cocoon: A structure formed by the clitellum that houses embryonic worms until they hatch.

coir: Coconut fiber or dust, a waste product of the coconut industry. Sold in compressed blocks as a worm bedding, it has high water-holding capacity. Used as substitute for peat moss.

compactor-transfer station: A facility that accepts solid waste and compacts it prior to transfer to a landfill or other refuse disposal facility.

compost: The biological reduction of organic waste to humus. Used to refer to both the process and the end product. One composts leaves, manure, and garden residues to obtain compost, which enhances soil texture and fertility when used in gardens.

consumer: An organism that feeds on other plants or animals.

culture: To grow organisms under defined conditions. Also, the product of such activity, as a bacterial culture.

cyst: A sac, usually spherical, surrounding an animal in a dormant state.

decomposer: An organism that breaks down cells of dead plants and animals into simpler substances.

decomposition: The process of breaking down complex materials into simpler substances. End products of much biological decomposition are carbon dioxide and water.

earthworm: A segmented worm of the phylum Annelida, over 5,000 species of which are terrestrial.

egg: A female sex cell capable of developing into an organism when fertilized by a sperm.

egg case: *See* cocoon.

***Eisenia andrei*:** The scientific name for the worm commonly used for vermicomposting. *Eisenia andrei* is a close relative of *Eisenia fetida*. It is entirely reddish, does not appear striped, and is sometimes known as the red tiger worm.

***Eisenia fetida* (formerly *Eisenia foetida*):** The scientific name for the most common redworm used for vermicomposting. It is characterized by lack of pigment between its reddish segments, thus showing a striping pattern. Common names include tiger worm, manure worm, and brandling.

enchytraeids: Small, white, segmented worms common in vermicomposting systems.

enzyme: A complex protein that provides a site for specific chemical reactions.

***Eudrilus eugeniae*:** The scientific name for a large worm of tropical origin commonly known as the African nightcrawler. Not suitable for vermicomposting in cold climates.

excrete: To separate and discharge waste.

feces: The waste discharged from the intestine through the anus. Manure.

fertilize: To supply nutrients to plants or to impregnate an egg.

genus: (pl.: genera) A category of classification grouping organisms with a set of characteristics more generalized than species characteristics.

girdle: *See* clitellum.

gizzard: A region in the anterior portion of the digestive tract whose muscular contractions help grind food.

green business: A business that prioritizes the environment in its practices and products. Typical policies include maximizing energy efficiency, avoiding pollution, and creating markets for recycled materials.

grit: Coarse or fine abrasive particles used by a worm to grind food in its gizzard.

hatchlings: Worms as they emerge from a cocoon.

heavy metals: Dense metals such as cadmium, lead, copper, and zinc that can be toxic in small concentrations. A buildup of heavy metals in garden soil should be avoided.

hemoglobin: An iron-containing compound in blood responsible for its oxygen-carrying capacity.

hermaphrodite: A term for an organism that possesses both male and female sex organs. Most earthworms are hermaphrodites (some are parthenogenetic, that is, have only female sex organs).

humus: A complex, highly stable material formed during the break-down of organic matter.

hybrid: Resulting from mating between individuals of two different species, normally producing sterile offspring, as when a horse-donkey breeding produces a mule. The term "hybrid redworm" is fairly common in the worm industry, but scientists do not accept this as proper usage of the term. No proof for hybridization among worm species exists.

hydrated lime: Calcium hydroxide. Do not use in worm bins. *See also* lime.

inoculate: To provide an initial set of organisms for a new culture.

Juniperus silicicola: Southern red cedar, considered by many botanists to be a variety of *Juniperus virginiana*.

Juniperus virginiana: Eastern red cedar, known by many names including aromatic cedar. It has a distinct and tell-tale scent.

leachate: Water that has run through a medium, causing soluble materials to dissolve and drain off.

leaf mold: Leaves in an advanced stage of decomposition.

lime: A calcium compound that helps reduce acidity in worm bins. Use calcium carbonate, ground limestone rock, eggshells, or oyster shells. Avoid caustic, slaked, and hydrated lime.

litter (leaf): Organic material on the forest floor containing leaves, twigs, decaying plants, and associated organisms.

Lumbricus rubellus: The scientific name for a worm species found in compost piles and soils rich in organic matter. Sometimes known as red marsh worm, dung worm, or redworm.

Lumbricus terrestris: The scientific name for a large burrow-dwelling nightcrawler. Known commonly as the Canadian nightcrawler in the United States, the dew worm in Canada.

macroorganism: An organism large enough to see by naked eye.

microbes: Very minute living things, whether plant or animal; bacteria, protozoa, fungi, actinomycetes.

microorganism: An organism requiring magnification for observation.

monoculture: Cultivation of a single species.

nematodes: Small (usually microscopic) roundworms with both free-living and parasitic forms. Not all nematodes are pests.

nightcrawler: A common name for *Lumbricus terrestris*, a large, burrow-inhabiting earthworm.

optimal: Most favorable conditions, such as for growth or for reproduction.

organic: Pertaining to or derived from living organisms.

overload: To deposit more food waste in a worm bin than can be processed aerobically.

pasteurize: To expose to heat long enough to destroy certain types of organisms.

pathogen: A disease-producing organism.

peat moss: Sphagnum moss that is mined from bogs, dried, ground, and used as an organic mulch. Although acidic, its light, fluffy texture and excellent moisture-retention characteristics make it a good medium for shipping worms. No longer recommended as a worm bedding because it is a limited resource and suitable alternatives exist.

***Perionyx excavatus*:** The scientific name for a tropical worm species found in India, southern parts of the United States, Australia, and elsewhere. One common name is Indian blue worm. Not suitable for vermicomposting in cold climates.

perlite: A lightweight volcanic glass used to increase aeration in potting mixtures.

pH: An expression for degree of acidity and alkalinity based upon the hydrogen ion concentration. The pH scale ranges from 0 to 14; pH 7 is neutral; less than 7, acid; greater than 7, alkaline.

pharynx: The muscular region of the digestive tract immediately posterior to a worm's mouth.

pit-run: *See* bed run.

population density: The number of specific organisms per unit area; for example, 1,000 worms per square foot.

posterior: Toward the rear, back, or tail.

potting soil: A medium for potting plants.

pot worms: *See* enchytraeids.

prostomium: The sensitive fleshy lobe protruding above an earthworm's mouth.

protein: A complex molecule containing carbon, hydrogen, oxygen, and nitrogen; a major constituent of meat. Worms are approximately 60 percent protein.

putrefaction: The anaerobic decomposition of organic matter, especially protein, characterized by disagreeable odors.

redworm: A common name for *Eisenia fetida, Eisenia andrei,* and also *Lumbricus rubellus. E. fetida* and *E. andrei* are the primary redworms used for vermicomposting.

regenerate: To replace lost parts.

run-of-pit: *See* bed run.

saddle: *See* clitellum.

salt: Salts are formed in worm bins as acids and bases combine, having been released from decomposition of complex compounds.

secrete: To release a substance that fulfills some function within the organism. Secretion of slime by a worm helps retain moisture and protect its body from injury by coarse soil particles.

segment: One of numerous disk-shaped portions of an earthworm's body bounded anteriorly and posteriorly by membranes.

seminal fluid: Fluid that contains sperm that are transferred to an earthworm's mate during copulation.

setae: Bristles on each segment used in locomotion.

sexually mature: Possessing a clitellum and capable of breeding.

side-dressing: The application of nutrients on the soil surface away from stems of plants.

slaked lime: Calcium hydroxide. Do not use in worm bins.

species: A basic category of biological classification, characterized by individuals that can breed together and produce offspring that can also produce young.

sperm: Male sex cells.

sperm-storage sacs: Pouches that hold sperm received during mating.

subsoil: Mineral-bearing soil located beneath humus-containing topsoil.

taxonomist: A scientist who specializes in classifying and naming organisms.

Thuja occidentalis: Northern white cedar is primarily used for fencing and posts. It is also used for cabin logs, lumber, poles and shingles. It has a distinctive moderate smell when worked.

Thuja plicata: Western red cedar, popular as a roofing material because it durable, a good insulator, and looks good. It has a strong aromatic smell when worked.

topdressing: Nutrient-containing materials placed on the soil surface around the base of plants.

toxic: Poisonous, life-threatening.

toxoplasmosis: Disease caused by the protozoan *Toxoplasma gondii*.

vermicast: A single worm casting or a quantity of worm castings. Worms "work" material by ingesting, excreting, and reingesting it. Vermicast is extensively worm-worked and reworked. It may be over-worked and has probably lost plant nutrients as compared to vermicompost. Vermicast has a fine, smooth texture, which may dry with a crust on the surface. *See also* worm casting.

vermicompost: A mixture of partially decomposed organic waste, bedding, and worm castings. Contains recognizable fragments of plant, food, or bedding material as well as cocoons, worms, and associated organisms. As a verb, to carry out composting with worms.

vermicomposting: The process of using worms and associated organisms to break down organic waste into material containing nutrients for plant growth.

vermiculite: Lightweight potting material produced through expansion of mica by means of heat.

vermiculture: The raising of earthworms under controlled conditions.

white worms: *See* enchytraeids.

worm bin: A container designed to accommodate a vermicomposting system.

worm castings: Undigested material, soil, and bacteria deposited through the anus. Worm manure. *See also* vermicast.

worm:waste ratio: When setting up a worm bin, the relationship between weight of worms and weight of food waste to be processed on a daily basis.

Bibliography

Appelhof, Mary. "Basement worm bins produce potting soil and reduce garbage." Kalamazoo, MI: Flowerfield Enterprises, 1973.

_____. "Composting your garbage with worms." Kalamazoo, MI: Kalamazoo Nature Center, 1981, 1979.

_____. "Household scale vermicomposting," in Workshop on the Role of Earthworms in the Stabilization of Organic Residues. Vol. I: Proceedings, compiled by Mary Appelhof. Kalamazoo, MI: Beech Leaf Press, 1981. pp. 232–240.

_____. "Vermicomposting on a household scale," in Soil Biology as Related to Land Use Practices, Proceedings of the International Colloquium on Soil Zoology, edited by Daniel Dindal. Washington, D.C.: U.S. EPA, 1980. pp. 157–160.

_____. Videomicroscopy of Live Earthworms, final report to National Science Foundation. Kalamazoo, MI: Flowerfield Enterprises, 1994. (SBIR Award No. III-936127)

_____. Winter Composting with Worms, final report to National Center for Appropriate Technology. Kalamazoo, MI: Kalamazoo Nature Center, 1979.

_____. "Worms — a safe, effective garbage disposal," *Organic Gardening and Farming*, 21(8): 1974; pp. 65–69.

_____. "Worms vs. high technology," *Creative Woman*, 4(1): 1980; pp. 23–28.

Appelhof, Mary, Katie Webster, and John Buckerfield. "Vermicomposting in Australia and New Zealand," *BioCycle*, 37: June 1996; pp. 63–66.

Appelhof, Mary, Michael Tenenbaum, and Randy Mock. "Energy considerations: resource recycling and energy recovery," presentation before the Resource Recovery Advisory Committee, South Central Michigan Planning Council. Kalamazoo, MI, July 1980.

Appelhof, Mary, Michael Tenenbaum, Randy Mock, Cheryl Poche, and Scott Geller. Biodegradable Solid Waste Conversion into Earthworm Castings, final report to National Science Foundation. Kalamazoo, MI: Flowerfield Enterprises, 1981. (ISP-8009755)

Ball, Ian R., and Ronald Slys. "Turbellaria: Tricladida: Terricola," in *Soil Biology Guide*, edited by Daniel L. Dindal. New York, NY: John Wiley and Sons, 1990.

Barrett, Thomas J. *Harnessing the Earthworm*. Boston, MA: Wedgwood Press, 1959, 1947.

Bhawalkar, Uday S. *Vermiculture Ecotechnology*. Pune, India: Bhawalkar Earthworm Research Institute, 1995.

Bogdanov, Peter. *Commercial Vermiculture: How to Build a Thriving Business in Redworms*. Merlin, OR: Vermico, 1996.

Brown, Amy. *Earthworms in New Zealand*. Auckland: Reed Publishing, 1995.

_____. *Earthworms Unlimited*. Maryborough, Victoria, Australia: Kangaroo Press Pty Ltd, 1994.

Buckerfield, J.C., and K.A. Webster. "Earthworms, mulching, soil moisture and grape yields," *Australian and New Zealand Wine Industry Journal*, 11(1): 1996; pp. 47–53.

_____. "Earthworms as indicators of sustainable production," in Proceedings Inaugural Ecological Economics Conference, November 19–23. Coffs Harbour, New South Wales, 1995. pp. 333–339.

Cooke, A. "The effects of fungi on food selection by *Lumbricus terrestris L.*," in *Earthworm Ecology*, edited by J.E. Satchell. Cambridge, England: Chapman and Hall, 1983. pp. 365–373.

Cresswell, G.C. "Coir dust — a viable alternative to peat?" Rydalmere, New South Wales, Australia: Biological and Chemical Research Institute. 1994.

Darwin, Charles. *The Formation of Vegetable Mould, through the Action of Worms, with Observations on their Habits*. New York: D. Appleton and Company, 1881, 1898.

Dindal, Daniel L. "Ecology of compost: a public involvement project." Syracuse, New York: NY State Council of Environmental Advisors and the State University of New York College of Environmental Science and Forestry, 1972.

_____. "Soil organisms and stabilizing wastes." *BioCycle, Journal of Waste Recycling*. 1978. 19(4); pp. 8–11.

_____. "The Decomposer Food Web." Script and 70 color slide set. The JG Press, Emmaus, PA, 1980.

Edwards, C.A., and Edward F. Neuhauser, editors. *Earthworms in Waste and Environmental Management*. The Hague, Netherlands: SPB Academic Publishing BV, 1988.

Edwards, C.A., and J.R. Lofty. *Biology of Earthworms*, 2nd edition. London, United Kingdom: Chapman and Hall, 1977.

Edwards, C.A., and P.J. Bohlen. *Biology and Ecology of Earthworms*, 3rd edition. London, United Kingdom: Chapman and Hall, 1996.

Ernst, David. *The Farmer's Earthworm Handbook: Managing Your Underground Money-Makers*. Brookfield, WI: Lessiter Publications, 1995.

Geller, E. Scott, Richard A. Winett, and Peter B. Everett. *Preserving the Environment*. Elmsford, NY: Pergamon, 1982.

Goldstein, Jerome. *Recycling*. New York, NY: Schocken Books, 1979.

Goldstein, Nora. "The state of garbage in America," *BioCycle*, 38: 1997; pp. 62–67.

Hambly, Sam. "The Allsaw insulated composter." Downsview, Ontario: Camp Allsaw, date unknown.

Handreck, Kevin Arthur. "Earthworms for gardeners and fishermen." Adelaide, Australia: CSIRO Division of Soils, 1978.

Hartenstein, R., E.F. Neuhauser, and J. Collier. "Accumulation of heavy metals in the earthworm *Eisenia foetida*," *Journal of Environmental Quality*, 9: 1980; pp. 23–26.

Hartenstein, R., E.F. Neuhauser, and D.L. Kaplan. "Reproductive potential of the earthworm *Eisenia foetida*," *Oecologia*, 43:1979; pp. 329–340.

Hendrix, Paul, editor. *Earthworm Ecology and Biogeography in North America*. Boca Raton, FL: Lewis Publishers, 1995.

Home, Farm and Garden Research Associates. *Let an Earthworm Be Your Garbage Man*. Eagle River, WI: Shields, 1954.

Houseman, Richard (revised by), Darryl Sanders. "Springtails." *extension .missouri.edu,* U.S. Department of Agriculture, June 2014, 7 April 2017.

Institute for Local Self-Reliance. Three discussion papers of the Grassroots Recycling Network. Internet, 1997.

James, Sam. Personal communication. Fairfield, IA: Maharishi University, 10 March 1997.

Kaplan, D.L., R. Hartenstein, and E.F. Neuhauser. "Coprophagic relations among the earthworms *Eisenia foetida*, *Eudrilus eugeniae* and *Amynthas* spp," *Pedobiologia*, 20: 1980; pp. 74–84.

Kaplan, D.L., E.F. Neuhauser, R. Hartenstein, and M.R. Malecki. "Physicochemical requirements in the environment of the earthworm *Eisenia foetida*," *Soil Biology and Biochemistry*, 12(1): 1980; pp. 347–352.

Kretzschmar, A., editor. ISEE 4: 4th International Symposium on Earthworm Ecology. Special Issue of *Soil Biology Biochemistry*, 24(12): 1992.

Martin, Deborah L., and Grace Gershuny, editors. *The Rodale Book of Composting*. Emmaus, PA: Rodale Press, 1992.

Martin, J.P., J.H. Black, and R.M. Hawthorne. "Earthworm biology and production." University of California Cooperative Extension, 1976. (Leaflet #2828)

McCormack, Jeffrey H. "A review of whitefly traps," *The IPM Practitioner,* 3(10): 1981.

Minnich, Jerry, and Marjorie Hunt. *The Rodale Guide to Composting*. Emmaus, PA: Rodale Press, 1979.

Mitchell, Myron J., Robert M. Mulligan, Roy Hartenstein, and Edward F. Neuhauser. "Conversion of sludges into 'topsoils' by earthworms," *Compost Science*: 1977 Jul/Aug; pp. 28–32.

Morgan, Charlie. *Earthworm Feeds and Feeding*, 6th edition. Eagle River, WI: Shields, 1972.

Munday, Vivian, and J. Benton Jones, Jr. "Worm castings: how good are they as a potting medium?" *Southern Florist and Nursery-man*, 94(2): 1981; pp. 21–23.

Neuhauser, E.F., D.L. Kaplan, M.R. Malecki, and R. Hartenstein. "Materials supporting weight gain by the earthworm *Eisenia foetida* in waste conversion systems," *Agricultural Wastes*, 2: 1980; pp. 43–60.

Neuhauser, E.F., R. Hartenstein, and D.L. Kaplan. "Second progress report on potential use of earthworms in sludge management," in *Proceedings of Eighth National Conference on Sludge Composting*. Silver Springs, MD: Information Transfer, Inc., 1979. pp. 238–241

Myers, Ruth. *A Worming We Did Go!* Elgin, IL: Shields, 1968.

Rao, B.R., I. Karuna Sagar, and J.V. Bhat. "*Enterobacter aerogenes* infection of *Hoplochaetella suctoria*," in *Earthworm Ecology*, edited by J.E. Satchell. Cambridge, England: Chapman and Hall, 1983. pp. 383–393.

Reinecke, A.J., S.A. Viljoen, and R.J. Saayman. "The suitability of *Eudrilus eugeniae, Perionyx excavatus* and *Eisenia fetida* (*Oligochaeta*) for vermicomposting in southern Africa in terms of their temperature requirements." *Soil Biology Biochemistry*, 24: 1992; pp. 1295–1307.

Reynolds, John W. *The Earthworms (Lumbricidae and Sparganophilidae) of Ontario*. Toronto, Canada: Royal Ontario Museum, 1977.

Rouelle, J. "Introduction of amoebae and *Rhizobium japonicum* into the gut of *Eisenia fetida* (*Sav.*) and *Lumbricus terrestris L.*," in *Earthworm Ecology*, edited by J.E. Satchell. Cambridge, England: Chapman and Hall, 1983. pp. 375–381.

Satchell, John E. "Earthworm microbiology," in *Earthworm Ecology: From Darwin to Vermiculture*, edited by J.E. Satchell. Cambridge, England: Chapman and Hall, 1983. pp. 351–364.

_____. "Earthworm evolution: Pangaea to production prototype," in Workshop on the Role of Earthworms in the Stabilization of Organic Residues. Vol. I: Proceedings, compiled by Mary Appelhof. Kalamazoo, MI: Beech Leaf Press, 1981. pp. 3–35.

_____. "Lumbricidae," in *Soil Biology*, edited by A. Burges and F. Raw. London and New York: Academic Press, 1967. pp. 259–322.

Schaller, Friedrich. *Soil Animals*. Ann Arbor, MI: University of Michigan Press, 1968.

Seldman, Neil N. "Recycling — history in the United States," in Encyclopedia of Energy Technology and the Environment, 1995. pp. 2352–2368.

Vick, Nicholas A. "Toxoplasmosis," in *Grinker's Neurology*, 7th edition. Springfield, IL: Charles C. Thomas Publishers, 1976.

White, Stephen. "A vermi-adventure to India!" *Worm Digest*, 15: 1997; pp. 1, 27, 30.

Worden, Diane D., editor. Workshop on the Role of Earthworms in the Stabilization of Organic Residues, Vol. II Bibliography. Kalamazoo, MI: Beech Leaf Press, 1981.

Bibliography for 35th Anniversary Edition by Joanne Olszewski

Chapter 3

"Arsenic CAS #7440-38-2." U.S. Department of Health and Human Services, 12 March 2015. Web, 20 Oct. 2016. www.atsdr.cdc.gov.

"Chromated Arsenicals (CCA)." U.S. Environmental Protection Agency. Web, 10 Oct. 2016. www.epa.gov.

Curwick, Christy. "Questions and Answers about Western Red Cedar and Asthma." *Logging and Sawmilling Journal*, 29 December 2004. Web, 20, Oct. 2016. www.forestnet.com.

Groenier, James "Scott" and Stan T. Lebow. "Types of Wood Preservatives." United States Department of Agriculture, U.S. Forest Service, n.d. Web, 11 Oct. 2016. www.fs.fed.us.

Livingston, Jean. "What's in That Pressure-Treated Wood?" U.S. Department of Agriculture, Forest Service. December 2005. Web, 20 Oct. 2016. www.fpl.fs.fed.us.

"Overview of Wood Preservative Chemicals." United States Environmental Protection Agency, May 2016. Web, 10 Oct. 2016. www.epa.gov.

"Specific Types of Wood Preservatives." NIC National Pesticide Information Center. Oregon State University and the U.S. Environmental Protection Agency, 18 December 2015, Web. 11 Oct. 2016.

"Table 2: Identification of Preservatives Listed for the Pressure Treatment of Southern Pine Wood Products." Southern Forest Products Association, 2010, Web. 11 Oct. 2016. www.southernpine.com.

"Volume I: Conifers." USDA Forest Service, 1 November 2004. Web, 20 Oct. 2016. www.na.fs.us.

Chapter 4

"Future Fibers: Coir." Food and Agriculture Organization of the United Nations, n.d. Web, 28 November 2016. fao.org.

Priesnitz, Wendy. "Does Peat Moss Have a Place in the Ecological Garden?" *Natural Life Magazine*, n.d. Web, 28 November 2016. www.life.ca /naturallife/0712/asknlpeat.html.

Richards, Davi. "Coir Is Sustainable Alternative to Peat Moss in the Garden." Oregon State University Extension Service, 2 June 2006. Web, 28 November 2016. extension.oregonstate.edu.

Santiestevan, Cristina. "Questioning Peat Moss." *Rodale's Organic Life*, 22 December 2010. Web, 28 November 2016. www.rodaleorganiclife.com.

Wright, Stephen. "Indonesia Bolsters Call to Halt Peat-Swamp Work." *Northwest Arkansas Democrat Gazette*, December 11, 2016, 6G.

Chapter 5

"Earthworm Ecology." Earthworm Society of Britain, n.d. Web, 8 December 2016. earthwormsoc.org.uk.

"Earthworm Ecological Groups." University of Minnesota, n.d. Web, 1 December 2016. www.greatlakeswormwatch.org.

Edwards, Clive A., Norman Q. Arancon, and Rhonda Sherman. *Vermiculture Technology*, Boca Raton: CRC, 2011.

Hale, Cindy. *Earthworms of the Great Lakes*, Duluth: Kollath + Stensaas, 2013.

Holdsworth, Andy, Cindy Hale and Lee Frelich. "Contain those Crawlers." Minnesota Department of Natural Resources. August 2014. Web, 2 December 2016. dnr.state.mn.us.

"Invasion of the Exotic Earthworms!" National Park Service, n.d. Web, 3 December 2016. www.nps.gov.

"Invasive Earthworms." Ontario Invading Species Awareness Program, 2012. Web, 2 December 2016. www.invadingspecies.com.

Rombke, Jorg, et al. "DNA barcoding of earthworms (*Eisenia fetida*/andreoi complex) from 28 ecotoxicological test laboratories." *Applied Soil Ecology*, Volume 104, August 2016, pp. 3–11.

Stewart, Amy, *The Earth Moved: On the Remarkable Achievements of Earthworms*, Chapel Hill: Algonquin, 2004.

"What Is an Invasive Species?" USDA National Agricultural Library, 24 May 2016. Web, 9 December 2016. www.invasivespeciesinfo.gov.

Williams, Bernadette and Colleen Robinson Klug. "There's a New Creepy-Crawly in Wisconsin." *Wisconsin Natural Resources Magazine*, June 2015. Web, 3 December 2016. dnr.wi.gov.

Chapter 8

Rama, Devi K. and Mary A. Lourdhu. "Vermicomposting of poultry feather using *Eisenia foetida* and Indigenous eathworm: A comparative study." *International Journal of Scientific Research*, Volume 2, Issue 10, October 2013, pp. 1–2.

Chapter 9

Pada, Hilary. "New Worm Bin Harvest Method." City Farmer, Canada's Office of Urban Agriculture, 22 October 1996. Web, 10 December 2016. cityfarmer.org.

Chapter 11

Choate, Paul M. and R.A. Dunn. "Featured Creatures, Entomology & Nematology, Land Planarians." University of Florida, November 2015. Web, 27 December 2016. www.entnemdept.ugl.edu.

"Invasive Species Compendium Datasheet report for *Bipalium kwense*." Centre for Agriculture and Biosciences International, 28 December 2016. Web, 28 December 2016. www.cabi.org/isc /datasheetreport?dsid=112705.

Rust, M.K. and D.H. Choe. "Pest Notes Ants." University of California Agriculture and Natural Resources. October 2012. Web, 28 December 2016. ipm.ucanr.edu.

Stokes, Amber N. et al. "Confirmation and Distribution of Tetrodotoxin for the First Time in Terrestrial Invertebrates: Two Terrestrial Flatworm Species (*Bipalium adventitium* and *Bipalium kewense*)." *PLOS ONE.* June 2014, Volume 9, Issue 6, p. 1.

Chapter 12

Arancon. N.Q. and Clive A. Edwards. "The utilization of vermicomposts in horticulture and agriculture." Ohio State University, 2007. Web, 7 October 2016. www.osu.edu.

Coleman, Eliot. "Soil Blocks." Chelseagreen.com, n.d. Web, 10 December 2016. www.chelseagreen.com.

Hay, Frank S. "Nematodes — the good, the bad and the ugly." American Phytopathological Society, 2017. Web, 30 December 2016. www.apsnet .org.

"Peat-free planting mix recipe with coconut coir." Root Simple, 1 March 2012. Web, 7 October 2016. www.rootsimple.com.

Sherman, Rhonda and Brandon Hopper. "Earthworm Castings as Plant Growth Media." North Carolina State Extension, 16 October 2016. Web, 30 December 2016. http://composting.ces.ncsu.edu.

Chapter 13

Daigneau, Elizabeth. "Curbside Composting Added to a Major City: Is it Yours?" *Governing the States and Localities*, February 2012. Web, 9 January 2017. www.governing.com.

"GiNRE-Waste Statement." GrassRoots Recycling Network, n.d. Web, 6 January 2017. www.grrn.org.

Liss, Gary. "What Is Zero Waste?" National Recycling Coalition, 8 August 2016. Web, 10 November 2016. www.nrcrecycles.org.

Mellino, Cole. "Find Out Which U.S. City Shames You into Composting." EcoWatch, 29 January 2015. Web, 10, November 2016. www.ecowatch .com.

"Portland 2012 Plastic Bag Ordinance No. 185737." The City of Portland Oregon, 7 November 2012. Web, 10 December 2016. www.portlandoregon.gov.

"Recycling in America in the Bin." *The Economist*, 22 April 2015. Web, 10 December 2016. www.economist.com.

Schultz, Jennifer and Brooke Oleen. "State Beverage Container Deposit Laws." National Conference of State Legislatures, 12 April 2016. Web, 10 December 2016. www.ncsl.org.

Scott, Ryan. "The Bottom Line of Corporate Good." Forbes, 14 September 2012. Web, 20 November 2016. www.forbes.com.

"State Plastic and Paper Bag Legislation." National Conference of State Legislatures, 11 November 2016. Web, 10 December 2016. www.ncsl.org.

"There is no away." Zero Waste San Diego, 28 March 2016. Web, 4 January 2017. www.zerowastesandiego.org.

"Turning Food Waste into Energy at the East Bay Municipal Utility District." U.S. Environmental Protection Agency, 23 February 2016. Web, 10 January 2017. www.epa.gov.

"Why Is the City of Portland Expanding the Ban on Plastic Bags?" The City of Portland Oregon, n.d. Web, 10 December 2016. www.portlandoregon .gov.

Wilson, Amber. "Rotary members, church help supply 'have nots' with medical necessities." Medical Supplies Network, Inc., 4 January 2006. Web, 10 December 2016. www.msni.org.

Zanolli, Ashley. "Sustainable Food Management in Action." BioCycle, March 2012, Vol. 53, No.3, page 48. www.biocycle.net.

"Zero Waste Business Principles." GrassRoots Recycling Network, n.d. Web, 6 January 2017. www.grrn.org.

Acknowledgments

I would like to thank the following for their support in the revision of this third edition:

Peggy Konert, for being my sounding board and support person, and for always making sure I was fed well and got to the gym during this process.

My brother, Adam Olszewski, for being my sidekick when I needed to travel for this endeavor, and for being willing to go in my place when I could not.

Alice Beetz, for her archives on composting with worms. We miss you, Alice.

The Carroll and Madison Public Library Foundation for holding the Books and Bloom Literary Festival. Attending that event in Eureka Springs, Arkansas, was a catalyst in the production of this third edition.

Amy Stewart for her guidance and generosity. Thank you.

A very special thank-you to Mary Appelhof and Mary Frances Fenton, without whom none of this would have happened. It was their wisdom and devotion that produced this book, and their love and caring that placed it in my hands.

— Joanne Olszewski

Resources

ATTRA Sustainable Agriculture
https://attra.ncat.org
Farming resources

Biocycle
www.biocycle.net
Recycling resources

Center for EcoTechnology
http://www.cetonline.org
413-445-4556
Information and resources

City Farmer, Canada's Office of Urban Agriculture
http://www.cityfarmer.org
/wormcomp61.html
Information and resources

Cornell University
http://cwmi.css.cornell.edu
Information and resources

DriloBASE
http://taxo.drilobase.org
A database of earthworm information

Great Lakes Worm Watch
www.greatlakeswormwatch.org
Information and research teams

Instructables
www.instructables.com/howto
/worm+bin/
Worm bin construction, including continuous flow bins

National Conference of State Legislatures
www.ncsl.org/research
/environment-and
-natural-resources.aspx
Information on environmental laws, including plastic bag bans

NatureWatch
www.naturewatch.ca
Information on identifying worms

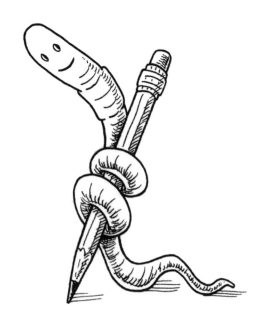

North Carolina State Vermicomposting
http://composting.ces.ncsu.edu
Information and resources

Rodale Institute
http://rodaleinstitute.org
610-683-1400
Classes, instructions, and information

Tilth Alliance
http://www.tilthalliance.org
206-633-0451
Classes, instructions, and information

University of Nebraska-Lincoln
http://lancaster.unl.edu/pest
/worms.shtml
Information and resources

Urban Ore
http://urbanore.com
510-841-7283
Information and resources

U.S. Composting Council
http://compostingcouncil.org
301-897-2715
Resources and programs

Washington State University
http://whatcom.wsu.edu/ag
/compost/easywormbin.htm
Instructions and information, including worm bin construction

Worm Culture
https://wormculture.org
Resources and information, including worm bin construction

Worm Woman Inc.
www.wormwoman.com
The original bin patented by Mary Appelhof in 1991; an aerated single bin made of recycled plastic, excellent for home use, demonstrations, and classroom activities

Zero Waste International Alliance
http://zwia.org
Information and resources

Index

Page numbers in *italic* indicate illustrations.

per pound of young redworms, 61
reaching maturity, 59
Buckerfield, John, 158
burying food waste, *17*, *81*, 81–82
buying worms
bed-run worms, *64*, 64–65
figuring how many you need,
60–63
free-range worms, 65
mail order, 71–72
sources of redworms, 63–65

C

calcium carbonate, 40–41, 76, 109
Canadian nightcrawler (*Lumbricus
terrestris*), 52–53, *53*, 162
castings (worm manure), 2–3, 6–9,
165
chemistry, 135–37
compost *vs.* castings, 6, 131
harvesting, 31, 93, 96–98, 101
in potting mixes, 139
in soil blocks, 139–40, *140*
sterilizing or not, 137
as topdressing, 141
toxic excess, 60, 92
weight of, 157–58
castings tea, 90
cats and toxoplasmosis, 80, 128
cautions
allergies, 129
avoid citrus, 77
avoid manure from dewormed animals, 36
avoid pressure-treated lumber, 19
avoid slaked lime, 41
don't feed worms salty food, 79
suitability of castings for plants,
135–36
use chemical-free containers, 18
cedar patio bench worm bin, 23,
28–29
suitability of cedar and other aromatic woods, 29

centipedes (class Chilopoda), 36,
115, 116, *119*, 119–20
checklist, worm composter process,
vii
coconut fiber or coir, 160
as bedding, 37–38, *38*
potting mixes, 139
soak for bedding use, 70
water-holding capacity, 38
cocoons, worm
formation and development, *57*,
57–60
mites that harm, 122, *122*
saving cocoons when harvesting,
95
temperature needs, 12
compost *vs.* castings, 6
continuous flow systems, 27, *27*

D

Darwin, Charles, 51, 108, 158, 167
dead worms
buildup of toxic castings, 8
extreme cold or heat, 102–5, *105*
food used up, 6
lack of oxygen, 16, 71
lifespan of a worm, 111
other possible causes, 111
population density, 60, 65
survival in your garden, 134
Dindal, Dr. Daniel, 78, 114, 116, 167
disease organisms, 128–29

E

earthworm types and species,
40–53
African nightcrawler (*Eudrilus
eugeniae*), 48, *48*
anecic, 44, *47*
Asian jumping worms (genus
Amynthus), 49–50, *50*
Canadian nightcrawler, 52–53, *53*
common North American species,
52

CULTIVATE YOUR GREEN THUMB
with More Books from Storey

The Complete Compost Gardening Guide
by Barbara Pleasant & Deborah L. Martin

This thorough, informative guide to materials and innovative techniques helps you turn an average vegetable plot into a rich incubator of healthy produce and an average flower bed into a rich tapestry of bountiful blooms all season long.

~~~~~~

The Complete Houseplant Survival Manual
by Barbara Pleasant

Keep your plants alive and thriving! Profiles of 160 flowering and foliage houseplants cover troubleshooting, care, pruning, repotting, and propagating. And whether you raise exciting imports or traditional favorites, the A-to-Z primer on houseplant health is an invaluable resource.

~~~~~~

Don't Throw It, Grow It!
by Deborah Peterson & Millicent Selsam

Rescue pits, seeds, tubers, and roots from your compost bin and grow lush houseplants. With step-by-step instructions, plant charts, and illustrations, you can turn kitchen scraps into beautiful, productive plants.

~~~~~~

Let It Rot! 3rd Edition
by Stu Campbell

Stop bagging leaves, grass, and kitchen scraps and turn household waste into the gardener's gold: compost. This classic guide offers accessible advice for starting and maintaining a composting system, building bins, and using your finished compost.

~~~~~~

Join the conversation. Share your experience with this book, learn more about Storey Publishing's authors, and read original essays and book excerpts at storey.com. Look for our books wherever quality books are sold or call 800-441-5700.